Excel
Basic Skills

Year 5
Ages 10-11

Mathematics

Get the Results You Want!

Damon James

Contents

Introduction

The aim of the ***Excel* Basic Skills Mathematics** series is to build on and reinforce students' basic skills in Mathematics. The books in the series support the requirements of Australian Curriculum Mathematics at each year level.

The ***Excel* Basic Skills Mathematics** series consists of seven books, one for each year level, from Kindergarten/Foundation to Year 6. The series is supported by other books in the ***Excel* Basic Skills** and **Advanced Skills** series.

Structure of the book

This book contains:

- thirty carefully graded double-page units of teaching and learning activities.
  - **Unit A** covers most of the **Number and Algebra** strand of the syllabus.
  - **Unit B** covers the rest of the **Number and Algebra** strand as well as the remaining **Measurement and Space** and **Statistics and Probability** strands of the syllabus.
- four double-page **revision units.**
- four four-page **NAPLAN-style tests.**

How to use this book

- Students should complete one unit per week. A suggested plan would be to complete the Unit A page for the week on one day and the Unit B page on another day of the same week.
- At the end of a sequence of units students should undertake the applicable Revision units. If students find particular revision questions difficult, they should revisit those areas in the previous sequence of units.
- After appropriate revision activities students should undertake the NAPLAN-style Test for those units. The revision work and testing should be completed on different days.

How to use this book with the *Excel* Basic Skills English series

For a complete **weekly English and Mathematics program** use this book in conjunction with the ***Excel* Basic Skills English Year 5** book. This way a student will have work set for four days a week—two days for English and two days for Mathematics.

How to assess students' progress

- The results of the work undertaken in each unit can be recorded on the marking grids. The marking grids on pages 6 and 7 (please see the example on page 4) are an easy-to-use diagnostic tool that indicates where students' strengths and weaknesses lie in relation to specific areas of Mathematics. These results can be used to gather extra information about students' progress and their further revision needs.

The *Excel* Basic Skills and Advanced Skills series

If students are experiencing difficulty, require additional practice or need extension in any area of the course, further books are available to support them in the ***Excel* Basic Skills** and **Advanced Skills** series. (Please see the comprehensive list of ***Excel*** books on page 5.)

The *Excel* step-by-step improvement plan

Step 1

Read the introduction on page 3.

Step 2

Read this page, along with the marking grids on pages 6 and 7.

- **Unit A and B Marking Grids**
  The results of the work undertaken in each unit can be recorded on the marking grids.
  These are an easy-to-use diagnostic tool that indicates where each student's strengths and weaknesses are in relation to specific areas of Mathematics.
  These results can be used to gather extra information about each student's progress and their further revision needs.
  For example, see the Sample Marking grid below.
- If a student is consistently getting more than one in five questions wrong in any area they need help in this area.
- When marking answers on the grid, simply mark incorrect answers with 'X' in the appropriate box. This will result in a graphical representation of areas needing further work. An example has been done above for the first five units. If a question has several parts, it should be counted as wrong if one or more mistakes are made.
- Remember that you can identify what type of question a student is having difficulty with in a topic. For example, in the grid below the student is having difficulty with multiplication questions.

	Addition	Subtraction	Multiplication	Multiplication	Division	Division	Addition	Addition	Subtraction	Subtraction	Multiplication	Multiplication	Division	Division	Place Value	Place Value	Money	Money	Decimals	Decimals	Rounding	Algebra
Question	1	2	3	4	5	6	7	8	9	10	11	12	13	14	15	16	17	18	19	20	21	22
Unit 1											X											
Unit 2												X										
Unit 3												X										
Unit 4											X	X										
Unit 5																						
Unit 6																						
Unit 7																						
Revision 1																						
Unit 8																						
Unit 9																						
Unit 10																						

This grid indicates that the student needs extra help and practice in multiplication questions.

Step 3

Refer to page 5: ***Excel* books to help you *get the results you want!***

- Under each topic there is a comprehensive list of books in our range to help a student.
  For example, if a student wants help with multiplication questions the books shown below will help them.
  Each ***Excel*** book has a comprehensive contents page that will identify the appropriate pages in the book to target the specific topic area that is causing problems.

Multiplication and Division

Excel **Basic Skills**

Mental Maths Strategies — 9781741251821

Multiplication and Division 5–6 — 9781864412895

Excel **Basic Skills Core**

English and Mathematics — 9781864412765

Excel **Advanced Skills**

Start Up Maths — 9781741252620

Excel **NAPLAN-style Tests**

Year 5 NAPLAN*-style Numeracy Tests — 9781741253603

Year 5 NAPLAN*-style Tests — 9781741121739

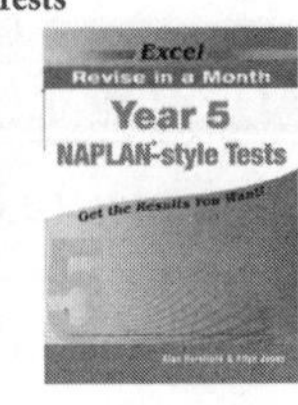

9781741252088

Excel books to help you *get the results you want!*

Number and Algebra

Place Value, Algebra and Operations

Excel **Basic Skills**

9781741251821

Excel **Basic Skills Core**

9781864412765

Excel **Advanced Skills**

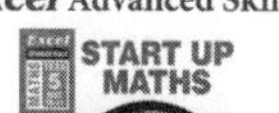

9781741252620

Excel **NAPLAN-style Tests**

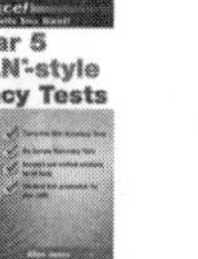

9781741253603

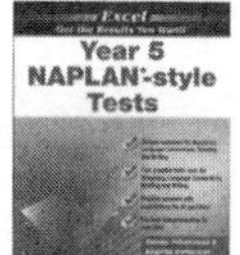

978174121739

9781741252088

Addition and Subtraction

Excel **Basic Skills**

9781741251821

9781864412871

9781740200516

Excel **Advanced Skills**

9781741252620

Excel **NAPLAN-style Tests**

9781741253603

978174121739

9781741252088

Multiplication and Division

Excel **Basic Skills**

9781741251821

9781864412895

Excel **Basic Skills Core**

9781864412765

Excel **Advanced Skills**

9781741252620

Excel **NAPLAN-style Tests**

9781741253603

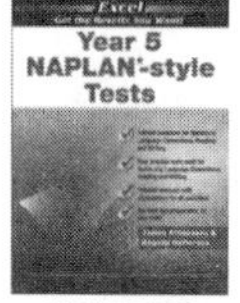

978174121739

9781741252088

Fractions, Decimals, Rounding and Money

Excel **Basic Skills**

9781741255904

9781864412901

9781741251821

Excel **Basic Skills Core**

9781864412765

Excel **Advanced Skills**

9781741252620

Excel **NAPLAN-style Tests**

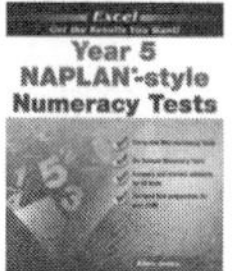

9781741253603

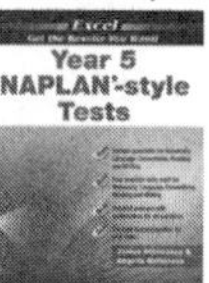

978174121739

9781741252088

Measurement and Space

Time and Measurement

Excel **Basic Skills**

9781741255904

9781741251821

Excel **Basic Skills Core**

9781864412765

Excel **Advanced Skills**

9781741252620

Excel **NAPLAN-style Tests**

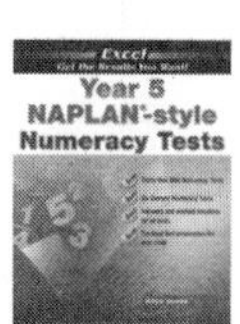

9781741253603

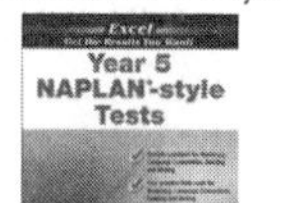

978174121739

9781741252088

2D/3D Shapes, Angles and Location

Excel **Basic Skills**

9781741251821

Excel **Advanced Skills**

9781741252620

Excel **NAPLAN-style Tests**

9781741253603

978174121739

9781741252088

Statistics and Probability

Graphs, Probability and Tables/Data

Excel **Basic Skills**

9781741251821

Excel **Advanced Skills**

9781741252620

Excel **NAPLAN-style Tests**

9781741253603

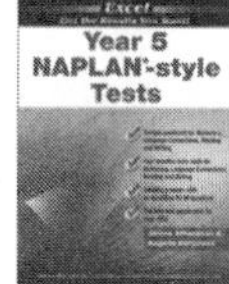

978174121739

9781741252088

Number and Algebra

	Addition	Subtraction	Multiplication	Multiplication	Division	Division	Addition	Addition	Subtraction	Subtraction	Multiplication	Multiplication	Division	Division	Place Value	Place Value	Money	Money	Decimals	Decimals	Rounding	Algebra
Question	**1**	**2**	**3**	**4**	**5**	**6**	**7**	**8**	**9**	**10**	**11**	**12**	**13**	**14**	**15**	**16**	**17**	**18**	**19**	**20**	**21**	**22**
Unit 1																						
Unit 2																						
Unit 3																						
Unit 4																						
Unit 5																						
Unit 6																						
Unit 7																						
Revision 1																						
Unit 8																						
Unit 9																						
Unit 10																						
Unit 11																						
Unit 12																						
Unit 13																						
Unit 14																						
Unit 15																						
Revision 2																						
Unit 16																						
Unit 17																						
Unit 18																						
Unit 19																						
Unit 20																						
Unit 21																						
Unit 22																						
Unit 23																						
Revision 3																						
Unit 24																						
Unit 25																						
Unit 26																						
Unit 27																						
Unit 28																						
Unit 29																						
Unit 30																						
Revision 4																						
Question	**1**	**2**	**3**	**4**	**5**	**6**	**7**	**8**	**9**	**10**	**11**	**12**	**13**	**14**	**15**	**16**	**17**	**18**	**19**	**20**	**21**	**22**

	Number and Algebra							Measurement and Space										Statistics and Probability				
	Addition	Subtraction	Multiplication	Division	Fractions	Fractions	Operations	Time	Time	Measurement—units	Measurement—length/area	Measurement—capacity/volume/mass	2D Shapes	3D Shapes	Location	Location	Angles	Probability	Probability	Tables/Data	Graphs	Graphs
Question	**1**	**2**	**3**	**4**	**5**	**6**	**7**	**8**	**9**	**10**	**11**	**12**	**13**	**14**	**15**	**16**	**17**	**18**	**19**	**20**	**21**	**22**
Unit 1																						
Unit 2																						
Unit 3																						
Unit 4																						
Unit 5																						
Unit 6																						
Unit 7																						
Revision 1																						
Unit 8																						
Unit 9																						
Unit 10																						
Unit 11																						
Unit 12																						
Unit 13																						
Unit 14																						
Unit 15																						
Revision 2																						
Unit 16																						
Unit 17																						
Unit 18																						
Unit 19																						
Unit 20																						
Unit 21																						
Unit 22																						
Unit 23																						
Revision 3																						
Unit 24																						
Unit 25																						
Unit 26																						
Unit 27																						
Unit 28																						
Unit 29																						
Unit 30																						
Revision 4																						
Question	**1**	**2**	**3**	**4**	**5**	**6**	**7**	**8**	**9**	**10**	**11**	**12**	**13**	**14**	**15**	**16**	**17**	**18**	**19**	**20**	**21**	**22**

1

+	12	8	16	19	21	15
8						

2

−	30	28	24	17	19	23
7						

3

×	1	3	7	6	8	4
5						

4

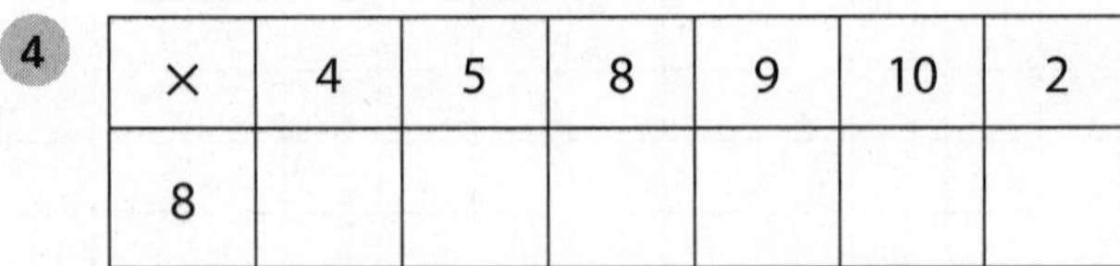

×	4	5	8	9	10	2
8						

5

÷	30	36	15	21	9	33
3						

6

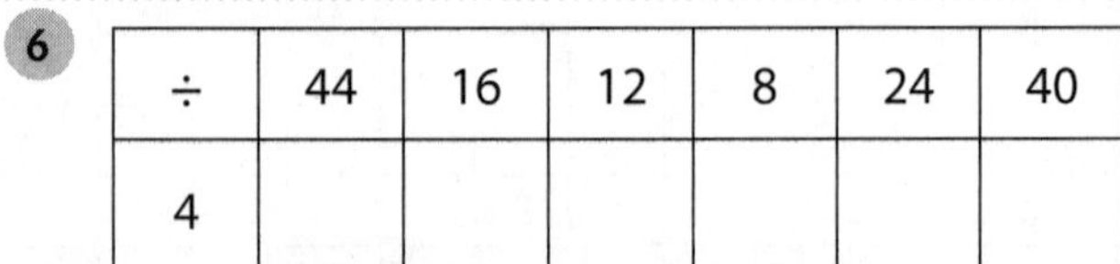

÷	44	16	12	8	24	40
4						

7

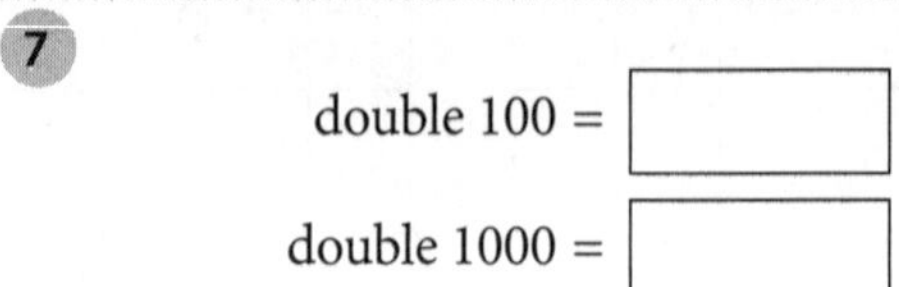

double 100 = ☐

double 1000 = ☐

8 Add 443 and 885.

☐

9 Complete the number line.

$246 - 113 =$ ☐

246

10 $171 - 56 =$ ☐

11

$$\begin{array}{r} 39 \\ \times \quad 6 \\ \hline \end{array}$$

12

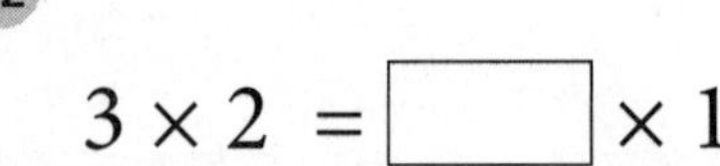

$3 \times 2 =$ ☐ $\times 1$

13

$$6\overline{)60}$$

14 66 balls were shared equally between 6 people. How many balls does each person have?

☐

15 Write 83 469 in words.

☐

16 Circle the largest number.

47 836   47 948

17 What is the change from $1.00 if the following is spent?

☐

18 What is the total number of  needed to make ?

☐

19 Write three-tenths as a decimal.

20 Locate 0.8 on the number line.

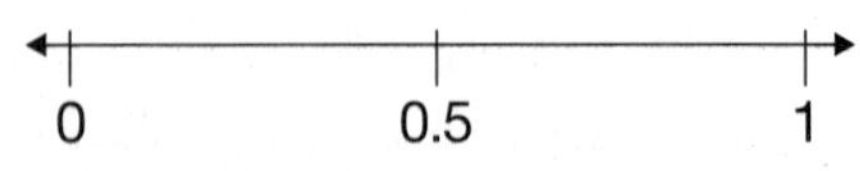

21 Round 683 to the nearest 10.

☐

22 $46 + 3 =$ ☐ $+ 19$

UNIT 1B

1 $200\,000 + 30\,000 =$ ______

2
$$\begin{array}{r} 29\,763 \\ -\ 14\,748 \\ \hline \end{array}$$

3
$$\begin{array}{r} 121 \\ \times\ \ 7 \\ \hline \end{array}$$

4 $5\overline{)575}$

5 What fraction is shaded?

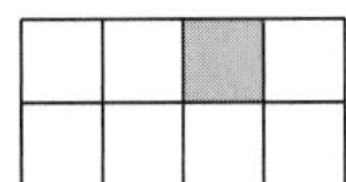

6 Correctly place $\frac{1}{2}$ on the number line.

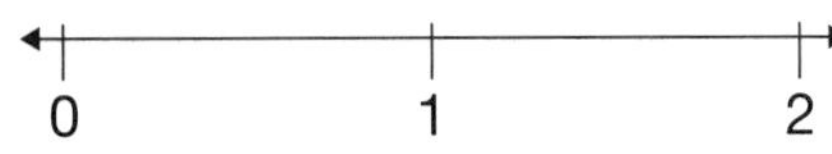

7 $59 - 14 + 6 - 3 =$ ______

8 Write the time in words.

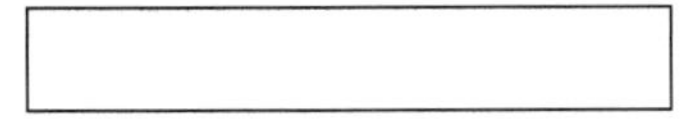

9 Write a quarter to seven on the digital clock.

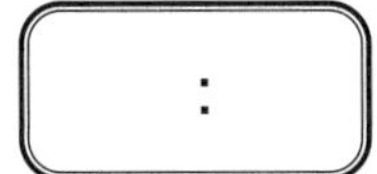

10 Circle the best unit to measure the length of the school oval.

g cm m L

11 Record the length marked on the ruler.

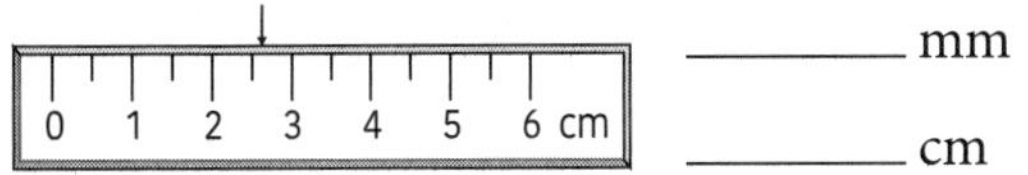

______ mm

______ cm

12 Colour the jug to show 1L.

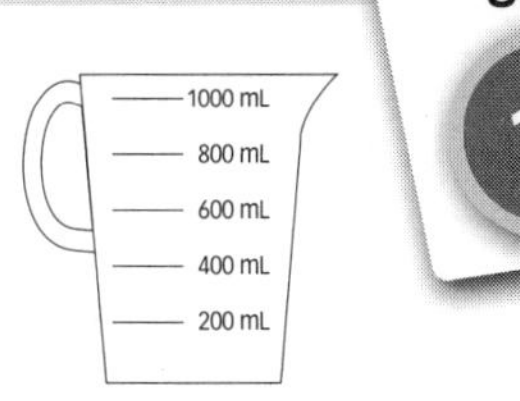

13 Name the shape.

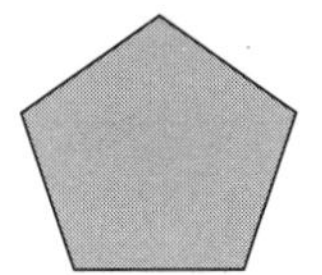

14 How many faces does this shape have?

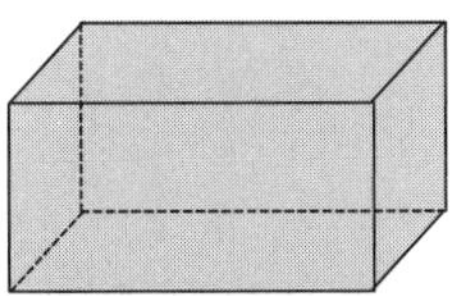

15 Which locker is below K?

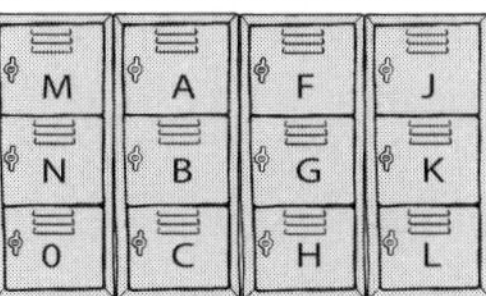

16 Darwin's coordinates are 3G. Label the dot for Darwin as D on the map.

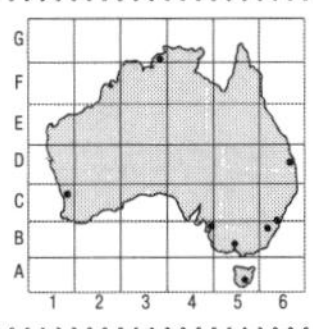

17 What is the number of interior angles?

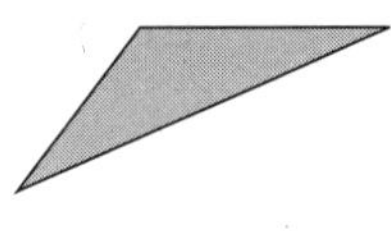

18 What is the chance you will play sport next week?

19 List the different outcomes when throwing one coin.

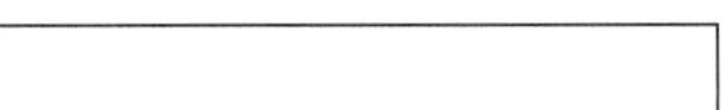

20 Write tally marks to show the number of apples.

29 apples 14 pears

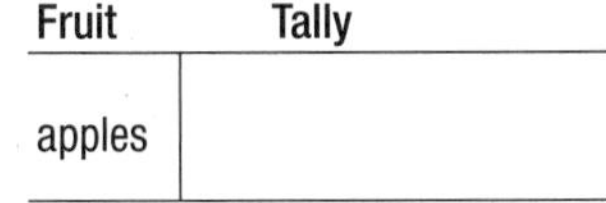

Fruit	Tally
apples	

21 What is the total number of boats?

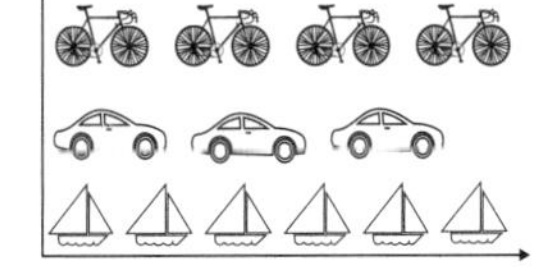

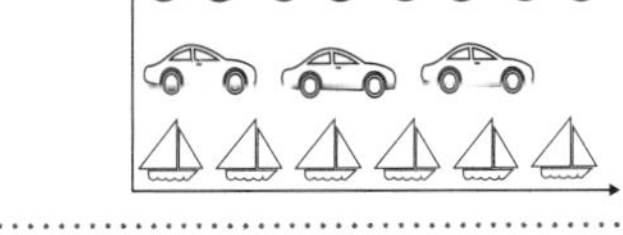

22 How far had Jack travelled in 4 seconds?

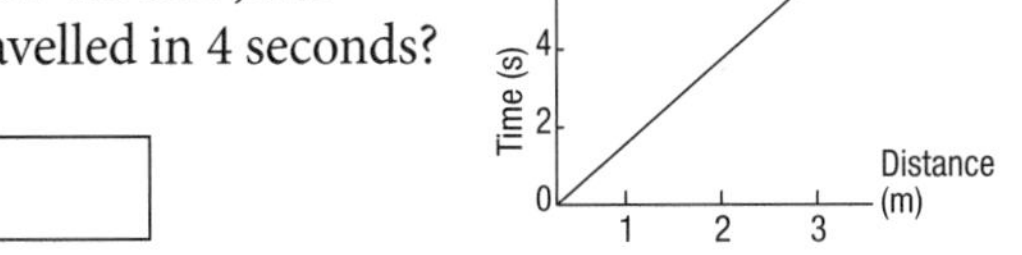

1

+	9	21	14	17	10	11
9						

2

–	30	24	17	12	21	28
8						

3

×	6	9	3	8	2	10
5						

4

×	4	5	1	9	7	0
8						

5

÷	12	9	18	21	30	3
3						

6

÷	16	40	8	4	28	32
4						

7

double 110 = ☐

double 1100 = ☐

8 Add 976 and 229.

☐

9 Complete the number line.

657 – 218 =

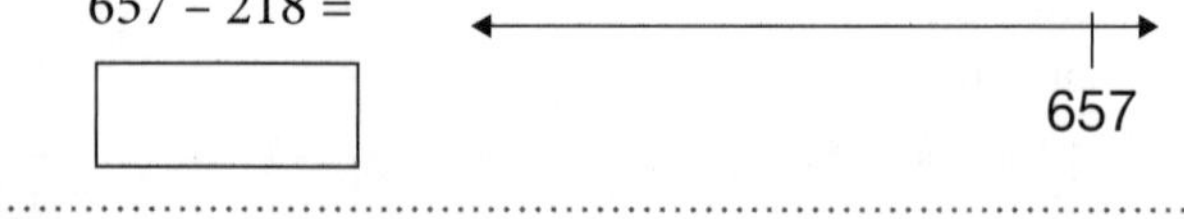

10

$$385 - 39 = \square$$

11

$$\begin{array}{r} 46 \\ \times\ \ 5 \\ \hline \end{array}$$

12

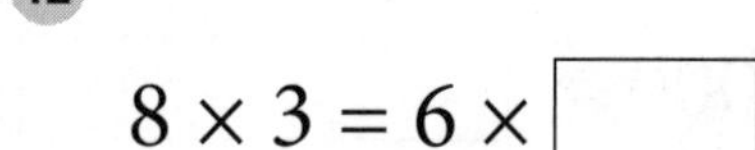

$$8 \times 3 = 6 \times \square$$

13

$$9\overline{)90}$$

14 45 apples were shared equally between 9 people. How many apples does each person have?

☐

15 Write 104 000 in words.

☐

16 Circle the largest number.

53 090 53 009

17 What is the change from $1.00 if the following is spent?

☐

18 What is the total number of 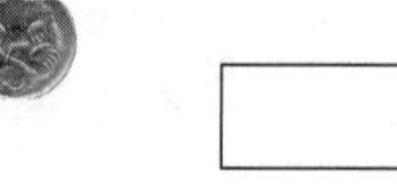 needed to make ?

☐

19 Write seven-tenths as a decimal.

20 Locate 0.2 on the number line.

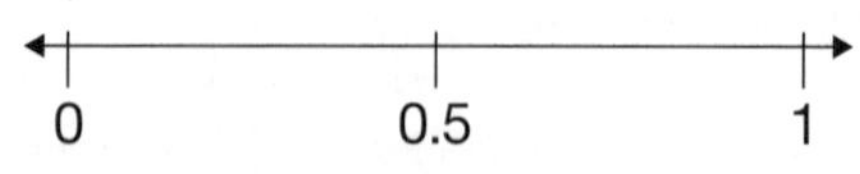

21 Round 418 to the nearest 10.

22

$$18 + 17 = \square + 20$$

1 $400\,000 + 70\,000 =$ ☐

2

$$\begin{array}{r} 38\,146 \\ -\ 27\,942 \\ \hline \end{array}$$

3

$$\begin{array}{r} 142 \\ \times\ \ 8 \\ \hline \end{array}$$

4

$$6\overline{)696}$$

5 What fraction is shaded?

6 Place $\frac{1}{4}$ in the correct location on the number line.

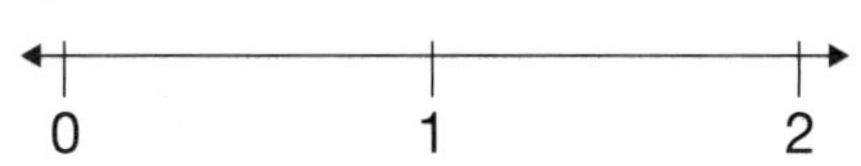

7 $76 + 13 - 26 + 9 =$ ☐

8 Write the time in words.

9 Write a quarter past three on the digital clock.

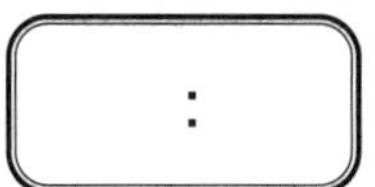

10 Circle the best unit to measure the weight of an elephant.

cm mL kg m

11 Record the length marked on the ruler.

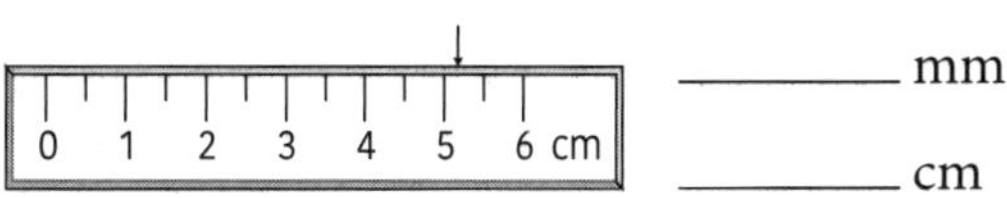

_______ mm

_______ cm

12 Colour the jug to show $\frac{1}{2}$ L.

13 Name the type of triangle.

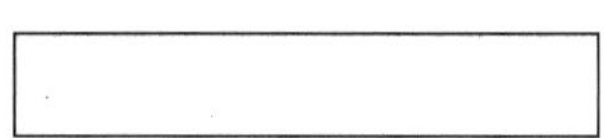

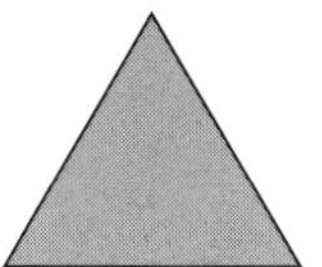

14 How many faces does this shape have?

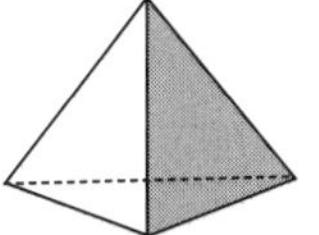

15 Which locker is next to N?

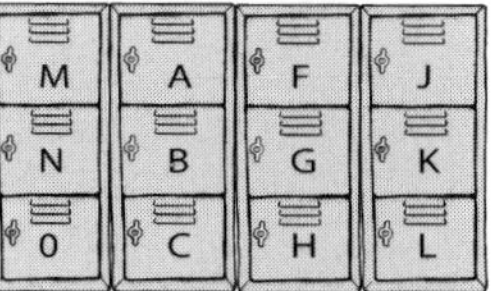

16 Adelaide's coordinates are 4B. Label the dot for Adelaide as A on the map.

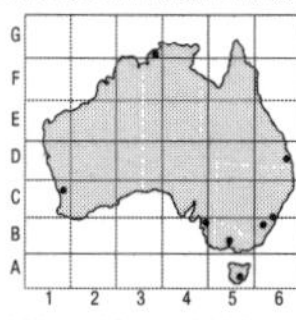

17 What is the number of interior angles?

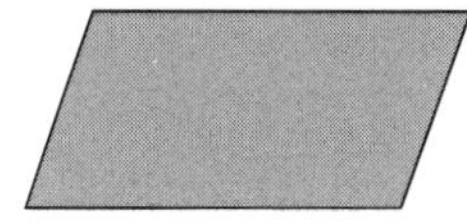

18 What is the chance it will snow tomorrow?

19 List the different outcomes when rolling a standard 6-sided dice.

20 Write tally marks to show the number of cats.

14 dogs 27 cats

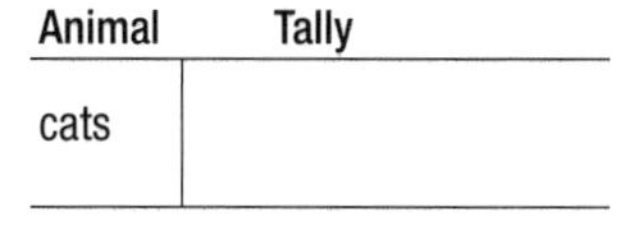

Animal	Tally
cats	

21 What is the total number of animals?

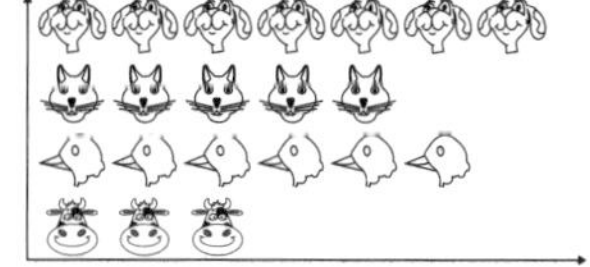

22 What was Ellie's height when she was 12 years old?

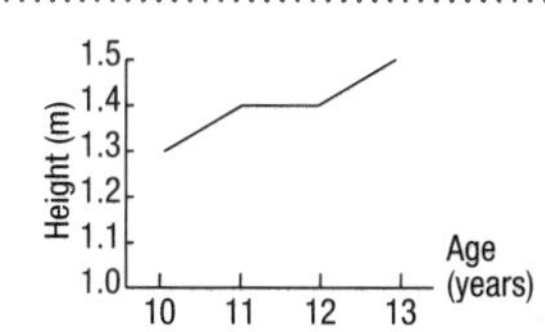

1

+	8	7	19	22	11	4
7						

2

–	9	21	12	14	18	30
6						

3

×	0	8	7	3	6	1
5						

4

×	2	5	8	10	4	9
8						

5

÷	33	12	3	15	21	9
3						

6

÷	12	40	8	16	28	44
4						

7

double 150 = ☐

double 1500 = ☐

8 Add 234 and 377.

☐

9 Complete the number line.

783 – 257 =

☐

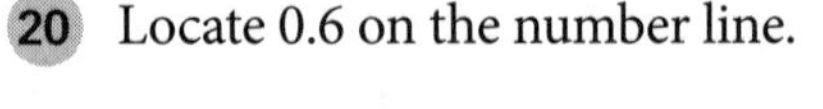

10 492 – 276 = ☐

11

$$\begin{array}{r} 72 \\ \times\ 7 \\ \hline \end{array}$$

12 $6 \times \square = 4 \times 9$

13 $7\overline{)70}$

14 28 oranges were shared equally between 4 boys. How many oranges does each boy have?

☐

15 Write 310 626 in words.

☐

16 Circle the largest number.

67 420 67 328

17 What is the change from $1.00 if the following is spent?

☐

18 What is the total number of  needed to make 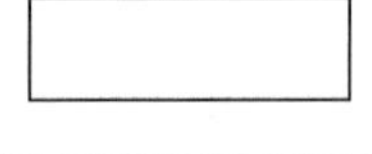 ?

☐

19 Write four-tenths as a decimal.

☐

20 Locate 0.6 on the number line.

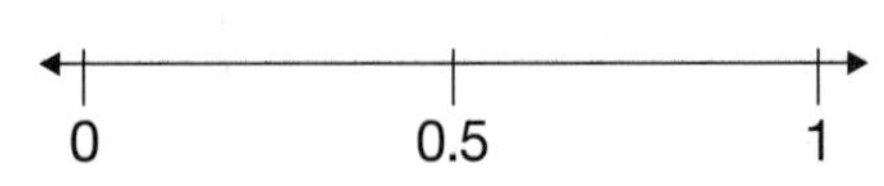

21 Round 769 to the nearest 10.

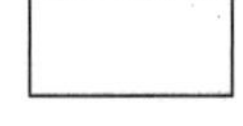

22 26 + 17 = 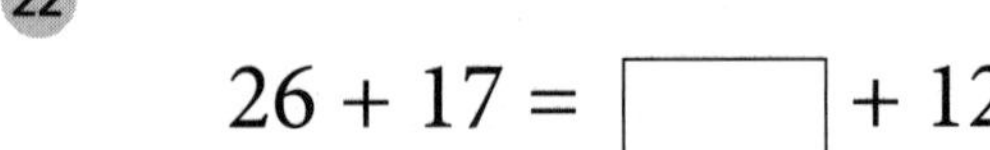 + 12

1 $700\,000 + 900\,000 = $ ☐

2

$$\begin{array}{r} 83\,916 \\ -\ 74\,326 \\ \hline \end{array}$$

3

$$\begin{array}{r} 149 \\ \times\quad 7 \\ \hline \end{array}$$

4 $4\overline{)768}$

5 What fraction is shaded?

6 Correctly place $\frac{1}{3}$ on the number line.

0 1 2

7 $100 - 62 + 35 - 7 = $ ☐

8 Write the time in words.

9 Write a quarter to six on the digital clock.

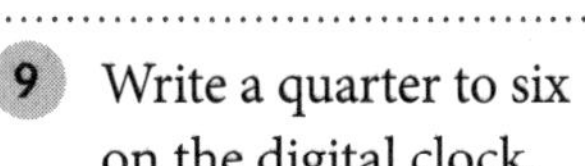

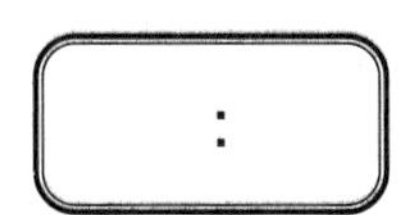

10 Circle the best unit to measure the height of a house.

L g mm m

11 Record the length marked on the ruler.

0 1 2 3 4 5 6 cm

______ mm

______ cm

12 Colour the jug to show 250 mL.

1000 mL
800 mL
600 mL
400 mL
200 mL

13 Name the shape.

14 How many faces does this shape have?

15 Which locker is above 8?

1	4	7	10	13
2	5	8	11	14
3	6	9	12	15

16 Brisbane's coordinates are 6D. Label the dot for Brisbane as B on the map.

17 What is the number of interior angles?

18 What is the chance you will fly in a plane next week?

19 List the different outcomes of spinning this spinner.

B R G

20 Write tally marks to show the number of flies.

47 flies 15 bees

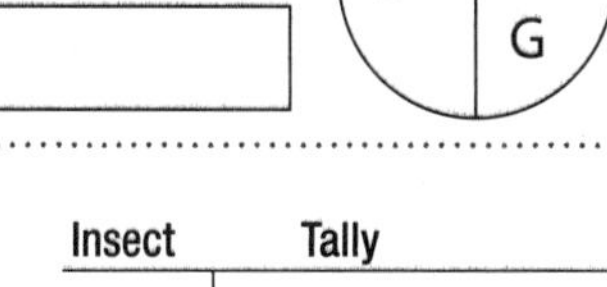

Insect	Tally
flies	

21 How many more apartments are there than houseboats?

22 At what time was it 15 °C?

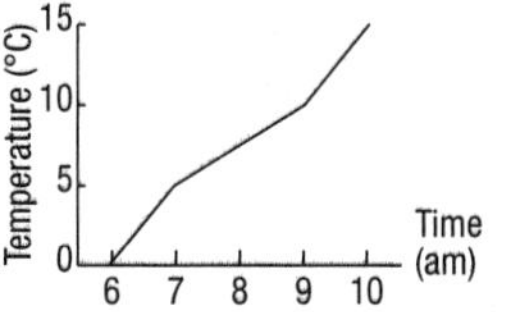

1

+	10	13	6	4	19	12
11						

2

–	30	18	25	10	21	27
9						

3

×	1	3	8	7	9	6
5						

4

×	2	5	10	4	0	7
8						

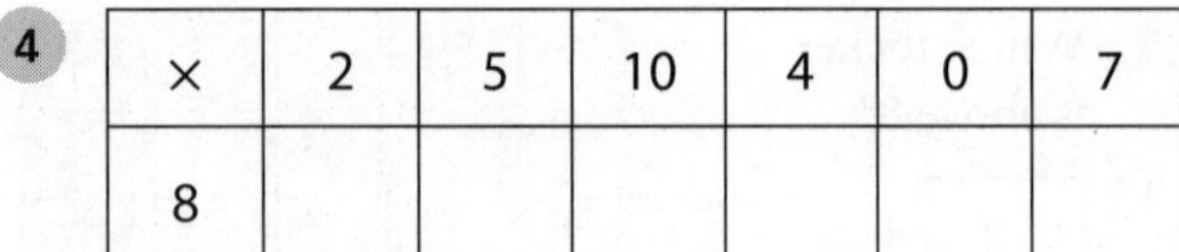

5

÷	33	12	9	18	6	15
3						

6

÷	24	40	8	36	16	20
4						

7 double 160 = ☐

double 1600 = ☐

8 Add 270 and 467.

☐

9 Complete the number line.

751 – 238 =

☐

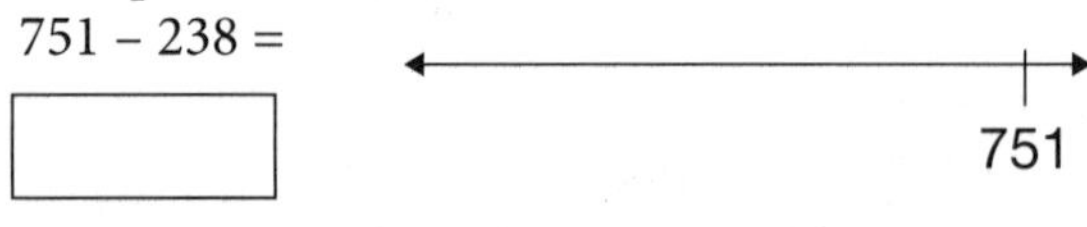

10 $683 - 147 =$ ☐

11

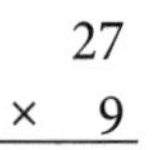

$$\begin{array}{r} 27 \\ \times \quad 9 \\ \hline \end{array}$$

12 $5 \times 8 = 4 \times$ ☐

13

$$8\overline{)80}$$

14 21 books were shared equally between 7 children. How many books does each child have?

☐

15 Write 409 301 in words.

☐

16 Circle the largest number.

32 460 31 095

17 What is the change from $1.00 if the following is spent?

☐

18 What is the total number of  needed to make $3.00?

☐

19 Write twelve-tenths as a decimal.

☐

20 Locate 0.1 on the number line.

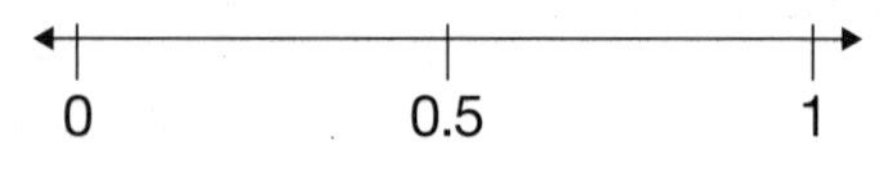

21 Round 901 to the nearest 10.

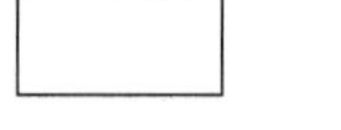

22 $23 + 18 =$ ☐ $+ 15$

1 $50\,000 + 700\,000 =$ ☐

2

$$\begin{array}{r} 47\,136 \\ -\ 29\,437 \\ \hline \end{array}$$

3

$$\begin{array}{r} 126 \\ \times\quad 9 \\ \hline \end{array}$$

4 $8\overline{)960}$

5 What fraction is shaded?

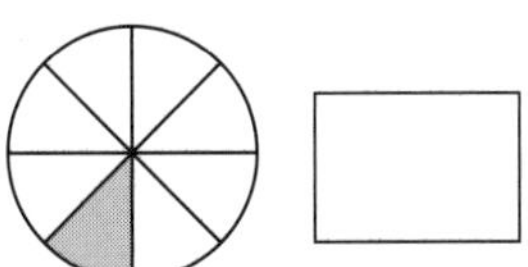

6 Correctly place $\frac{1}{10}$ on the number line.

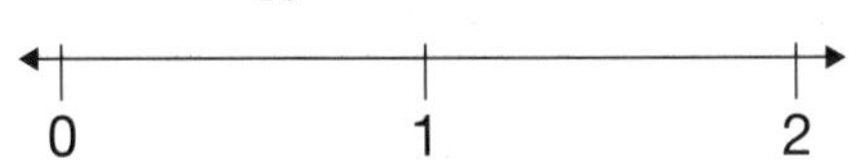

7 $106 - 19 + 37 - 74 =$ ☐

8 Write the time in words.

9 Write a quarter past twelve on the digital clock.

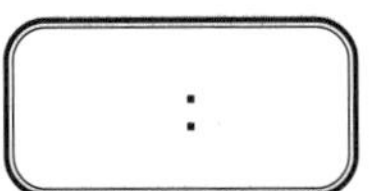

10 Circle the best unit to measure the volume of water in a bucket.

L kg m km

11 Record the length marked on the ruler.

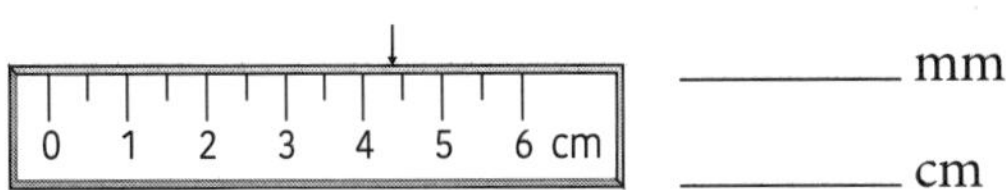

_______ mm

_______ cm

12 Colour the jug to show 700 mL.

13 Name the type of triangle.

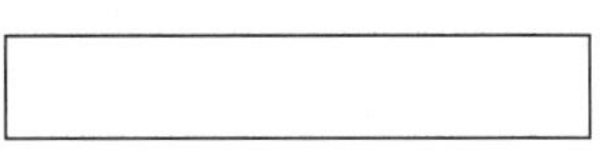

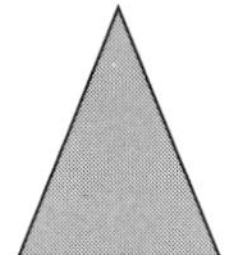

14 How many faces does this shape have?

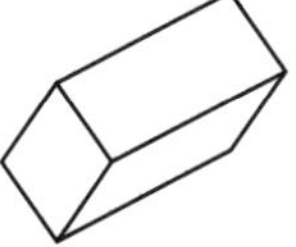

15 Which locker is below 2?

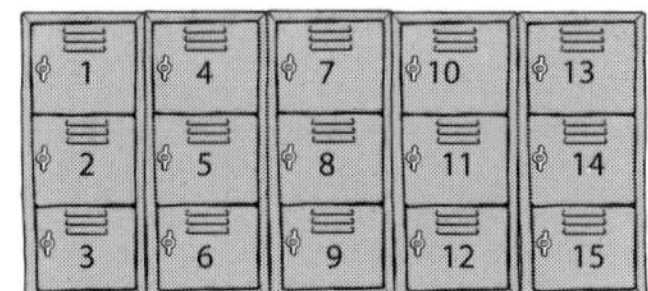

16 Perth's coordinates are 1C. Label the dot for Perth as P on the map.

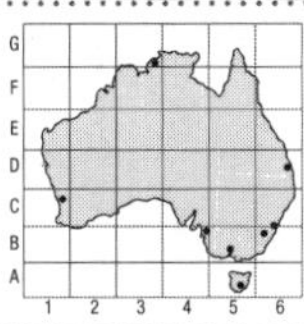

17 What is the number of interior angles?

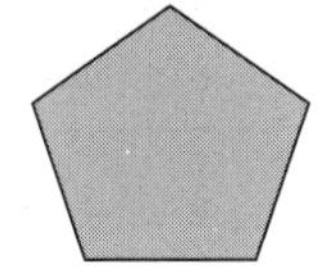

18 What is the chance my teacher next year will be a male?

19 List the different outcomes when throwing two coins.

20 Write tally marks to show the number of each colour pen.

Colour	Number	Tally
red	9	
blue	27	
green	6	

21 What type of ball did Jo lose the most of?

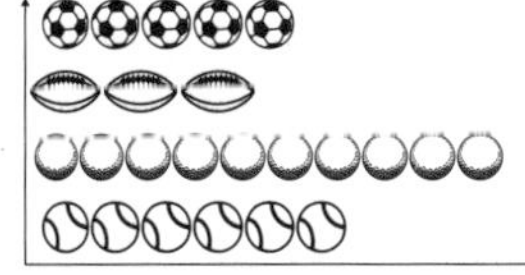

22 What was the amount of money made in March?

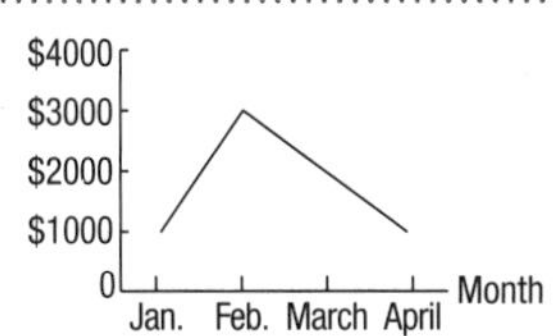

1

+	9	15	17	11	8	10
12						

2

−	27	16	18	24	21	30
11						

3

×	3	4	8	9	7	2
5						

4

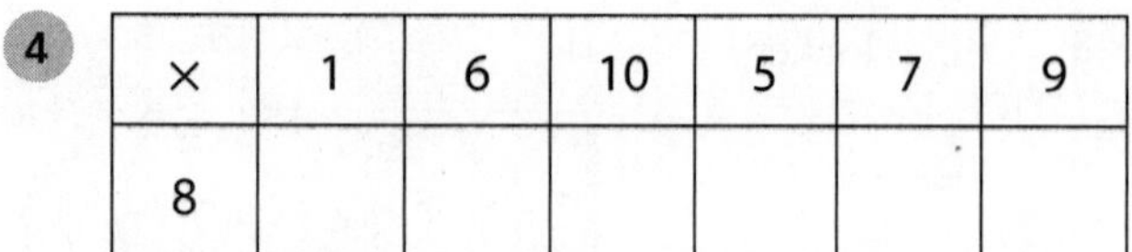

×	1	6	10	5	7	9
8						

5

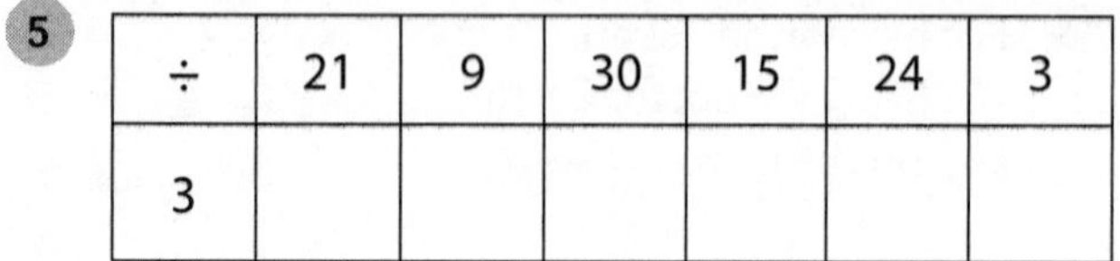

÷	21	9	30	15	24	3
3						

6

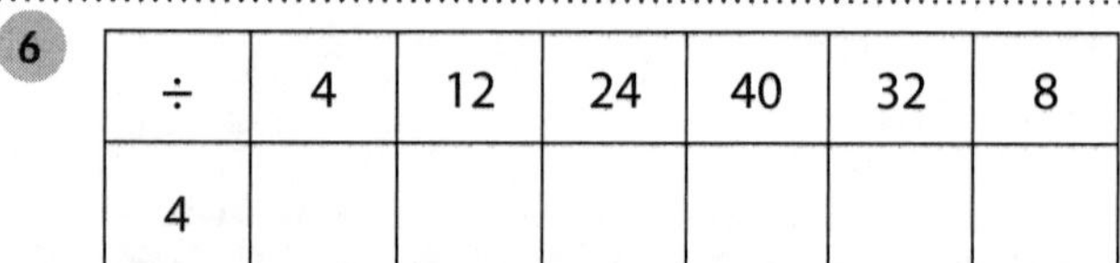

÷	4	12	24	40	32	8
4						

7

double 430 = ☐

double 4300 = ☐

8 Add 273 and 328.

☐

9 Complete the number line.

672 − 258 = ☐

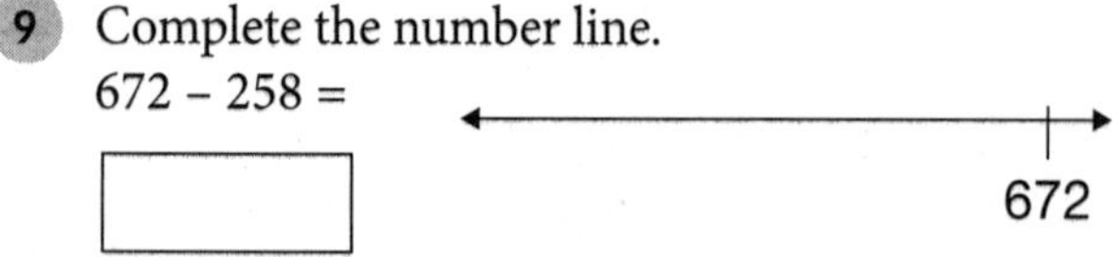

10

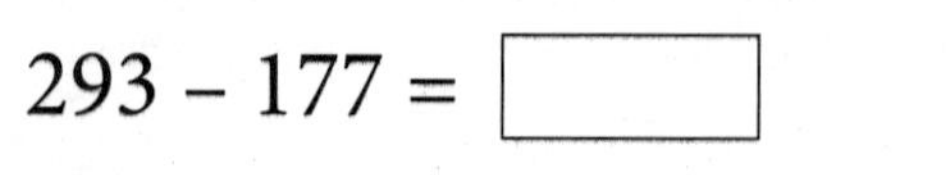

11

$$\begin{array}{r} 43 \\ \times \ \ 7 \\ \hline \end{array}$$

12

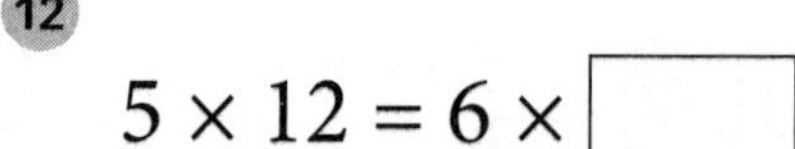

13

$$4\overline{)40}$$

14 56 pens were shared equally into groups of 7. How many pens are in each group?

☐

15 Write 117 039 in words.

☐

16 Circle the largest number.

324 116 324 204

17 What is the change from $1.00 if the following is spent?

☐

18 What is the total number of  needed to make [20c coin] ?

☐

19 Write twenty-one tenths as a decimal.

20 Locate 0.4 on the number line.

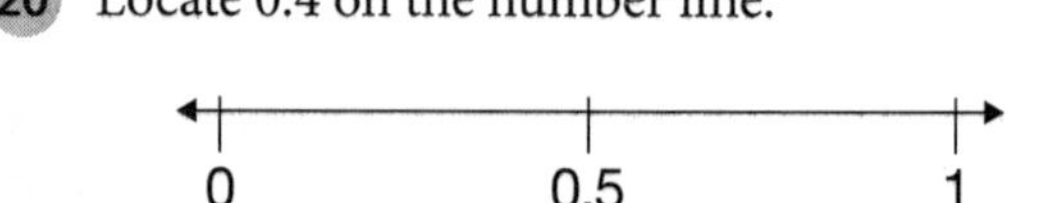

21 Round 311 to the nearest 10.

22

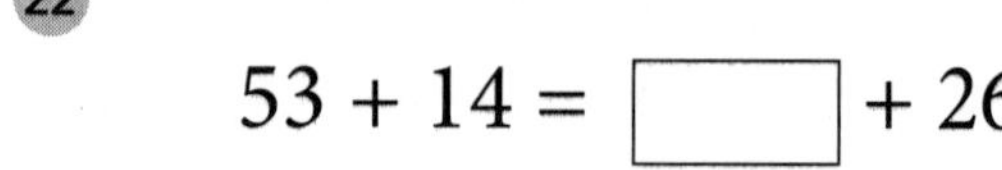

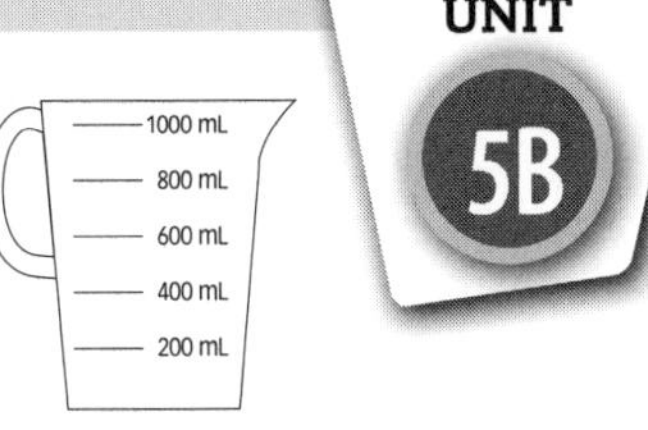

1

$8000 + 800\,000 =$ ☐

2

$$\begin{array}{r} 79\,634 \\ -\ 16\,479 \\ \hline \end{array}$$

3

$$\begin{array}{r} 276 \\ \times\quad 4 \\ \hline \end{array}$$

4

$$7\overline{)784}$$

5 What fraction is shaded?

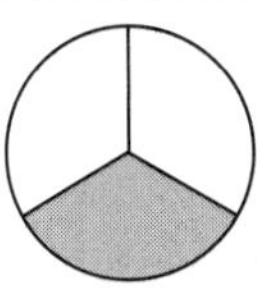

6 Correctly place $\frac{1}{8}$ on the number line.

7

$200 - 37 - 46 + 15 =$ ☐

8 Write the time in words.

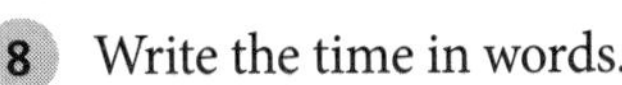

9 Write a quarter past nine on the digital clock.

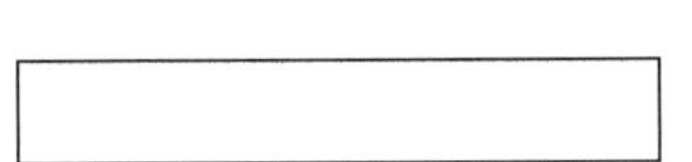

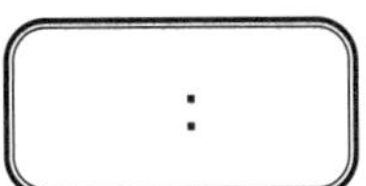

10 Circle the best unit to measure the mass of a pile of bricks.

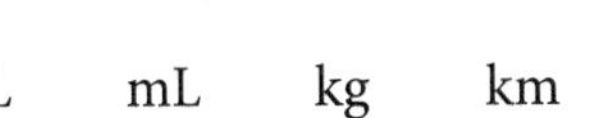

L mL kg km

11 Record the length marked on the ruler.

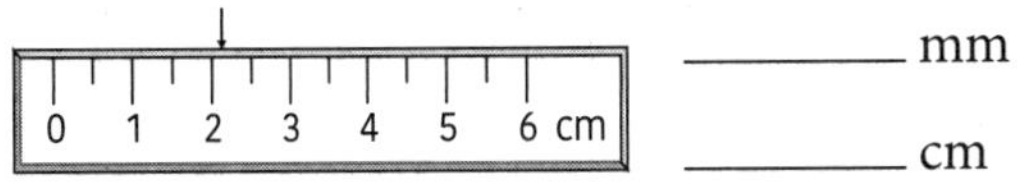

_________ mm

_________ cm

12 Colour the jug to show 300 mL.

13 Name the type of triangle.

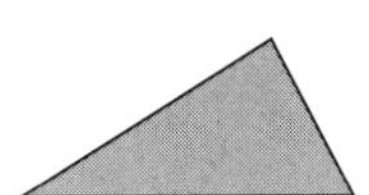

14 How many faces does this shape have?

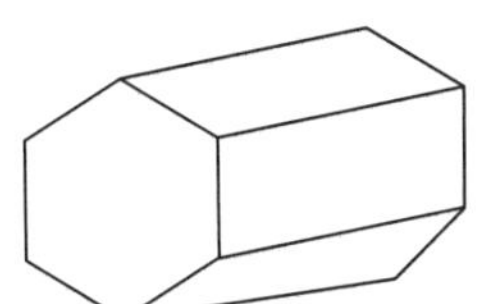

15 Which locker is in the middle?

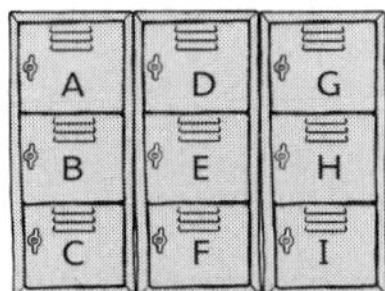

16 Hobart's coordinates are 5A. Label the dot for Hobart as H on the map.

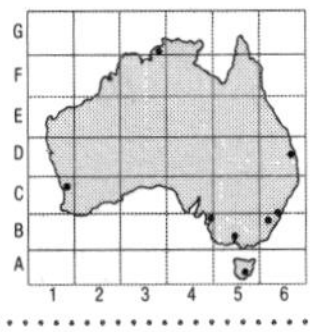

17 What is the number of interior angles?

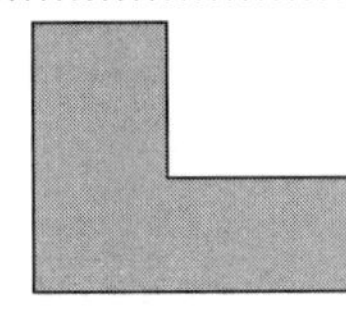

18 What is the chance you will grow up and become Prime Minister?

19 List the different outcomes from spinning this spinner.

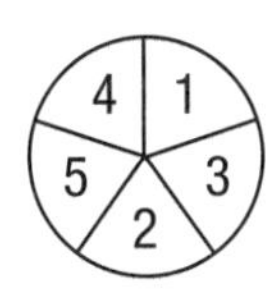

20 Write tally marks to show the number of each drink.

Drink	Number	Tally
juice	16	
milk	19	
water	4	

21 How many more sun than cloud stickers had Anita collected?

22 How many minutes did it take for the temperature of the water to reach 80 °C? ☐ min

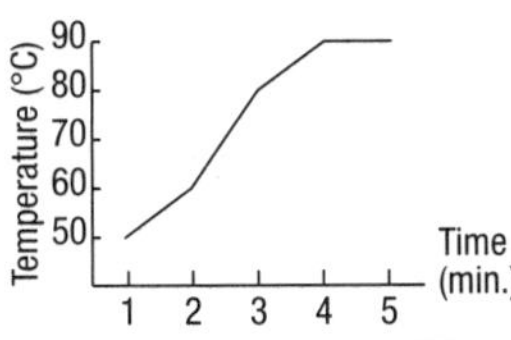

1

+	15	19	11	7	26	8
21						

2

–	40	27	33	18	49	25
15						

3

×	1	3	9	7	10	8
6						

4

×	11	6	0	4	7	12
7						

5

÷	20	60	40	15	55	25
5						

6

÷	88	24	48	80	32	72
8						

7 $800 + 600 = \square$

8 What is the total of 16, 40 and 27?

9

$$\begin{array}{r} 471 \\ -\ 247 \\ \hline \end{array}$$

10 $1800 - 1300 = \square$

11 5 groups of 15 = $\square$

12 $163 \times 8 = \square$

13 Circle the numbers divisible by 3.

30   28   35   15

14 $5\overline{)85}$

15 What is the value of the underlined digit?

32<u>8</u>16

16 Write two hundred and seventy-six thousand, four hundred and eight in digits.

17 What is the total cost of:

$4.35 and $6.54 ?

18 Zeb bought a chocolate bar for $1.10. What is Zeb's change from $2.00?

19 Circle the largest value.

0.63   0.36   0.06   0.66

20 Write 36 hundredths as a decimal.

21 Round 15 321 to the nearest hundred.

22 $30 \times \square = 300$

1

$$\begin{array}{r} 204\,638 \\ +\ 179\,642 \\ \hline \end{array}$$

2 What is the difference between 200 000 and 80 000?

3 There are 306 lettuce plants in one row. How many lettuce plants are in 7 rows?

4 True or false?

951 ÷ 8 = 112

5 What fraction of the group is shaded?

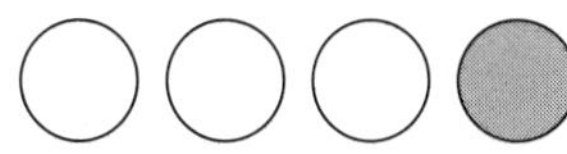

6 Colour each fraction to find the answer.

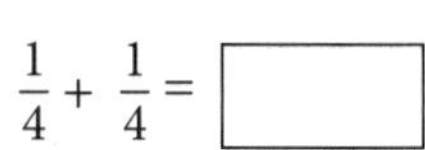

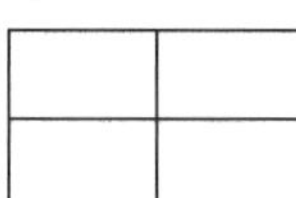

7

20 + 10 + 13 − 6 =

8 Draw a quarter past six on the clock.

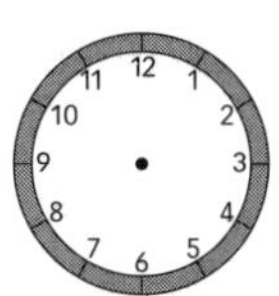

9 Write the time in words.

10

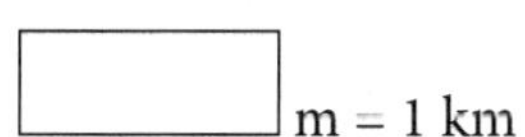

m = 1 km

11 Tick the most appropriate unit of measurement.

Object	mm	cm	m	km
length of a mobile phone				

12 What is the volume of the centicube model?

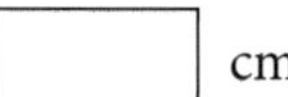

cm³

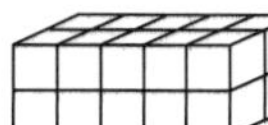

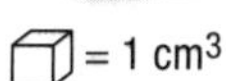

13 How many sides does this shape have?

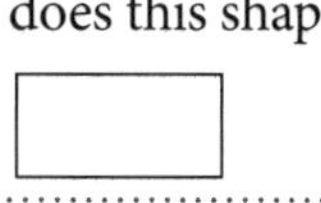

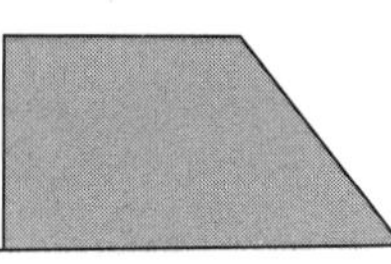

14 Name this 3D shape.

15 What is south of the pineapple?

16 What is at A3?

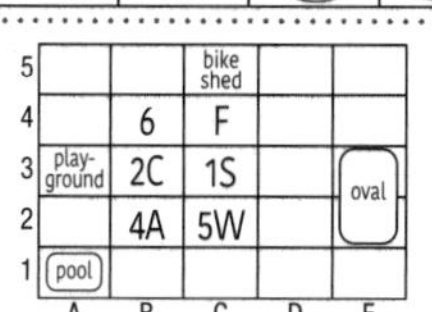

17 True or false? 360° is a right angle.

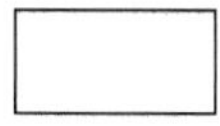

18 What is the most likely ball number to be selected from the bag?

19 Rate the likelihood that it will rain tomorrow on a scale of 0 (impossible) to 1 (certain).

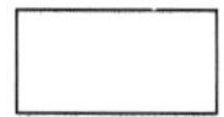

20 Complete the numbers on the tally chart.

Day	Tally	Number
Monday	𝍸 𝍸 ||	
Tuesday	𝍸 𝍸 𝍸	
Wednesday	𝍸 ||||	

21 How long did it take Jack to travel 3 m?

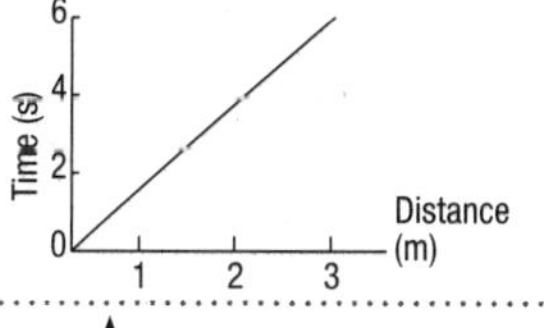

22 Draw a column graph showing:
8 cats
2 dogs
4 birds

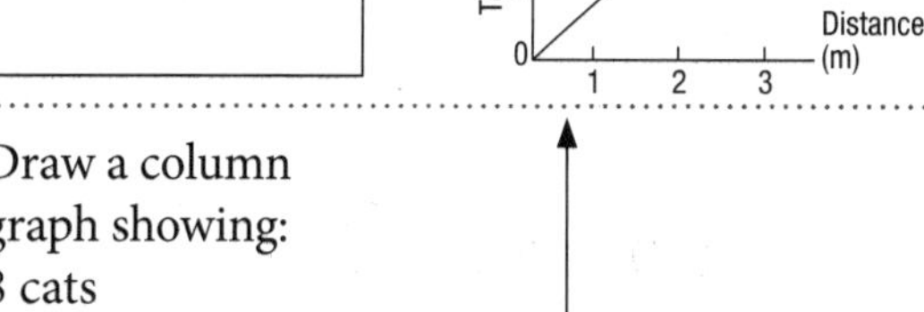

UNIT 7A

1

+	20	17	11	15	23	6
19						

2

–	21	18	30	36	29	47
11						

3

×	1	6	4	3	11	10
6						

4

×	11	1	9	12	4	3
7						

5

÷	30	50	10	15	45	5
5						

6

÷	8	64	24	80	48	16
8						

7 $700 + 200 = \square$

8 What is the total of 60, 80 and 23?

9

$$\begin{array}{r} 385 \\ -\ 266 \\ \hline \end{array}$$

10 $1400 - 240 = \square$

11 9 groups of 13 = ☐

12 $247 \times 6 = \square$

13 Circle the numbers divisible by 4.

22 36 44 18

14 $7\overline{)91}$

15 What is the value of the underlined digit?

610 4<u>3</u>7

16 Write three hundred and eight thousand, six hundred and ninety-four in digits.

17 What is the total cost of:

$8.56 and $7.92 ?

18 Jen bought a pencil for 85c. What was Jen's change from $2.00?

19 Circle the largest value.

0.92 0.29 0.96 0.30

20 Write 72 hundredths as a decimal.

21 Round 72 436 to the nearest hundred.

22 $14 \times 7 = \square$

UNIT 7B

1

$$\begin{array}{r} 796\,321 \\ +\ 483\,524 \\ \hline \end{array}$$

2 What is the difference between 70 000 and 900 000?

3 There are 6 tennis balls in a packet. How many balls are in 324 packets?

4 True or false?

322 ÷ 4 = 79

5 What fraction of the group is shaded?

6 Colour each fraction to find the answer.

$\frac{1}{5} + \frac{1}{5} =$

7

87 + 90 + 35 – 60 =

8 Draw a quarter to eleven on the clock.

9 Write the time in words.

1 : 30

10

cm = 1 m

11 Tick the most appropriate unit of measurement.

Object	mm	cm	m	km
distance from Sydney to Brisbane				

12 What is the volume of the centicube model?

cm³

= 1 cm³

13 How many sides does this shape have?

14 Name this 3D shape.

15 What is east of the kiwi fruit?

16 What is at C5?

5			bike shed		
4		6	F		
3	play-ground	2C	1S		oval
2		4A	5W		
1	pool				
	A	B	C	D	E

17 True or false? 30° is an acute angle.

18 What is the most likely colour ball to be selected from the bag?

R R G R G B G B

19 Rate the likelihood that you will watch television today on a scale of 0 (impossible) to 1 (certain).

20 Complete the numbers on the tally chart.

Month	Tally	Number
September	卌 III	
October	卌	
November	卌 卌 II	
December	卌 卌 I	

21 How old was Ellie when she was 1.3 m tall?

Height (m): 1.0, 1.1, 1.2, 1.3, 1.4, 1.5
Age (years): 10, 11, 12, 13

22 Draw a column graph showing:
9 cups
10 plates
4 bowls

1

+	21	19	7	15	10	38
11						

2

–	20	11	13	26	42	35
9						

3

×	4	6	3	8	9	1
8						

4

×	2	7	10	11	12	5
5						

5

÷	48	28	32	44	16	12
4						

6

÷	18	30	36	12	27	9
3						

7

double 600 = ☐

double 1200 = ☐

8 Greg added 27, 46 and 53. What was Greg's total?

☐

9 Complete the number line.

250 – 126 = ☐

250

10

900 – 600 = ☐

700 – 200 = ☐

11 Emily had 8 boxes of 10 pencils. How many pencils did Emily have?

☐

12

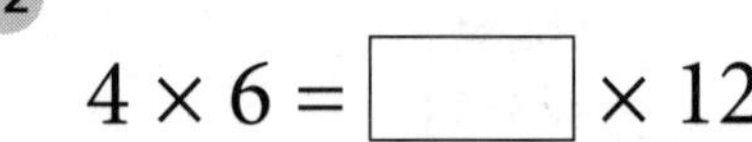

$4 \times 6 = \square \times 12$

13 Circle the numbers divisible by 6.

24 40 48 28 70

14 42 snacks are shared equally between 6 boys. How many snacks does each boy receive?

☐

15 Write 96 040 in words.

☐

16 Circle the largest number.

18 326 80 872 8998 81 005

17 Ruth spent 85c. How much change did she receive from $2.00?

☐

18 How many 20c make $5?

A 10 **B** 20
C 25 **D** 50

19 Write nine-tenths as a decimal.

20 Which is the location of 0.3 on the number line?

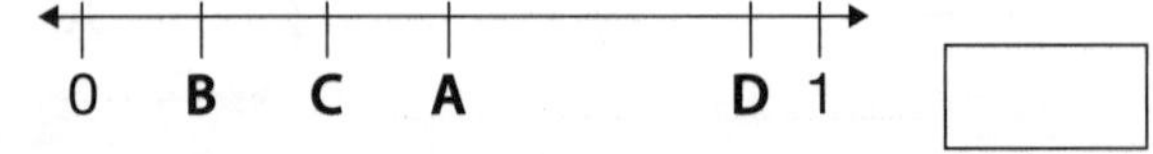

21 Round each number to the nearest hundred.

4783 ☐ 27 605 ☐ 72 093 ☐

22

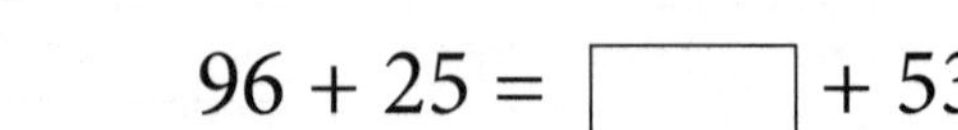

$96 + 25 = \square + 53$

REVISION

1B

1 What is the total of 328 149 and 384 666?

2

$$\begin{array}{r} 296\,108 \\ -\ \ 43\,742 \\ \hline \end{array}$$

3 Zara has 91 boxes with 6 pineapples in each box. How many pineapples does Zara have?

4 True or false?

648 ÷ 4 = 152

5 $\frac{1}{4}$ of the apples are eaten. What number of apples are eaten?

6 What is $\frac{1}{8} + \frac{6}{8}$?

7 90 + 60 – 27 + 6 =

8 What time is shown on the clock?

9 Match the digital time to the time on the clock.

A 6:00 **B** 12:00
C 11:30 **D** 12:30

10 Tick the most appropriate unit of measurement.

Object	mm	cm	m	km
length of school oval				
length of a pencil				

11 1 cm = ☐ mm

A 0.1 **B** 10
C 100 **D** 1000

12 What is the volume of the model?

☐ cm^3

= 1 cm^3

13 How many sides does a hexagon have?

14 How many faces does this object have?

15 What fruit is west of the banana?

16 What is at C2?

5			bike shed		
4		6	F		
3	play-ground	2C	1S		oval
2		4A	5W		
1	pool				
	A	B	C	D	E

17 Circle true or false for each.

True or false? 90° is a straight angle.
True or false? 110° is an obtuse angle.
True or false? 290° is an acute angle.

18 What is the chance of rolling a 5 or a 2 on a standard six-sided dice?

19 List the outcomes for this bag of balls.

G G B R R B R

20 How many goals were scored by Team A?

Team	Tally
A	𝍸 |||
B	𝍸 𝍸
C	𝍸

21 How many ice creams were sold on Tuesday?

Wednesday	🍦🍦🍦
Tuesday	🍦🍦🍦🍦🍦🍦
Monday	🍦🍦🍦🍦

22 What was the height of the bean plant at Week 3?

Height (cm): 10, 20, 30, 40, 50
Week: 1, 2, 3, 4, 5

1 Which equation represents the number line?

A 56 + 173 = 229
B 173 − 117 = 56
C 173 ÷ 2 = 56
D 173 − 119 = 56

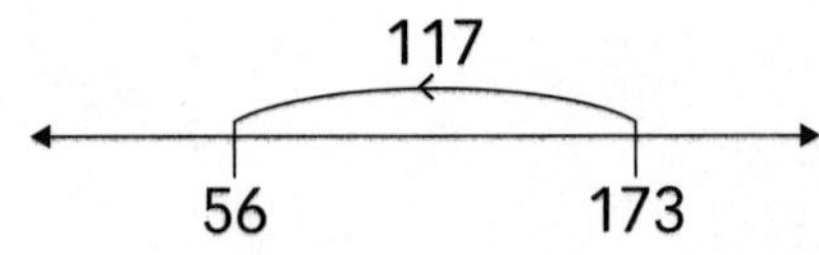

2 What is the missing number?

$$3 \times 8 = \square \times 4$$

A 6
B 7
C 8
D 10

3 Chris has 8 bags of carrots. In each bag there are 9 carrots.
How many carrots does Chris have altogether?

A 17
B 63
C 72
D 90

4 36 balls are placed equally into each of these containers.
How many balls are in each container?

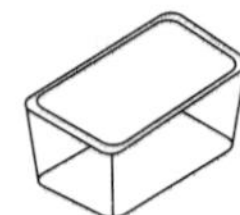 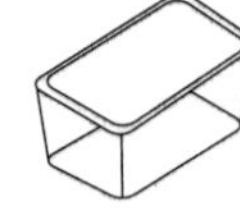 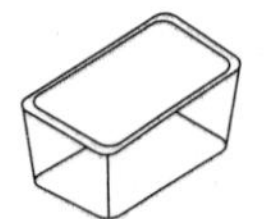 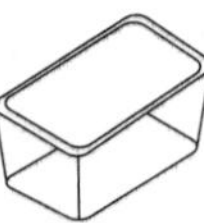

5 Which is the smallest number?

A 56 321
B 65 986
C 60 043
D 54 999

6 Ruby bought this apple with a $2 coin.
How much change did Ruby receive?

25c

A 75c
B 85c
C $1.75
D $1.85

7 Which position is 0.7 on the number line?

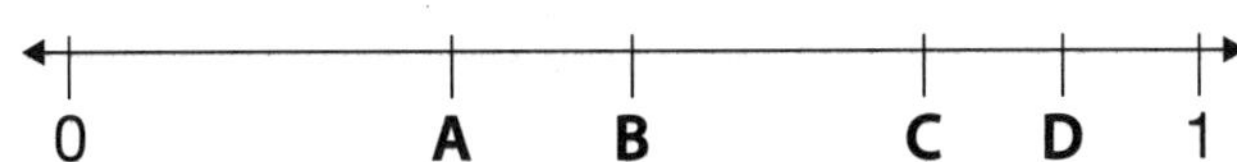

8 James rounds 1036 to the nearest hundred. What does James write?

A 1000
B 1100
C 1040
D 2000

9 32 618 + 43 842 =

A 76 450
B 76 460
C 75 450
D 75 460

10 What is $\frac{1}{3}$ of 12 oranges?

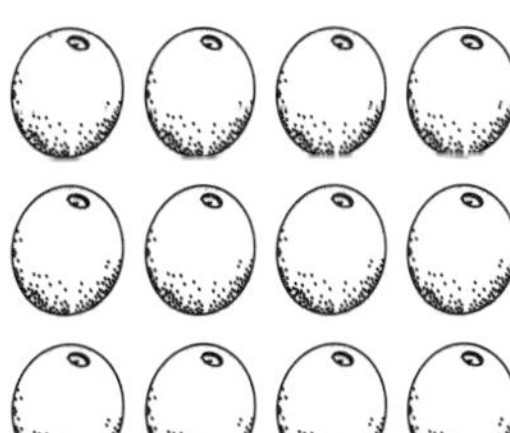

A 3

B 4
C 6
D 9

11 What time is shown on the clock?

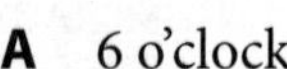

A 6 o'clock
B half past 12
C 12 o'clock
D half past 11

---

12 I am measuring the distance between Sydney and Brisbane.
Which is the best unit to use?

A mm
B m
C cm
D km

---

13 9 cm = ☐ mm

A 0.9
B 90
C 900
D 9000

---

14 How many faces does this box have?

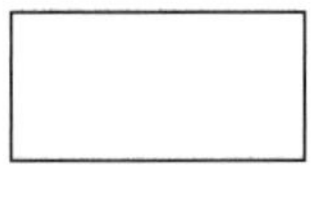

---

15 Jo is building a model.
What is the volume of Jo's model?

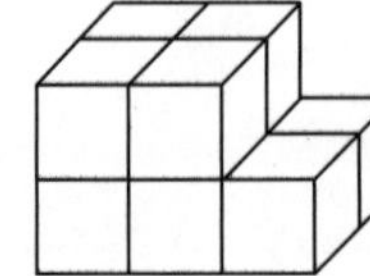

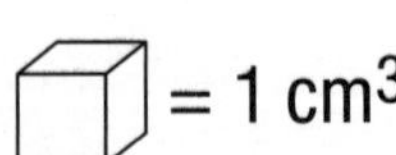

A 4 cm^3
B 5 cm^3
C 8 cm^3
D 10 cm^3

16 X marks the spot! What is to the south-west of the X?

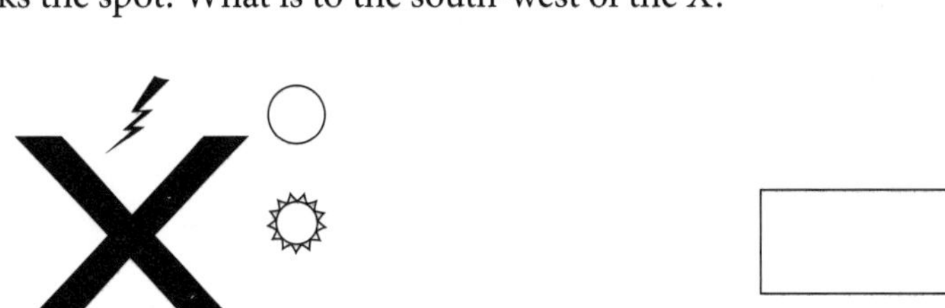

17 Which angle is obtuse?

A 30°
B 120°
C 300°
D 360°

18 These cards were placed into a bag.
Which number is most likely to be taken from the bag?

A 1
B 4
C 3
D 7

1	4	4
3	7	4

19 What is the value of the tally marks?

A 4
B 8
C 9
D 14

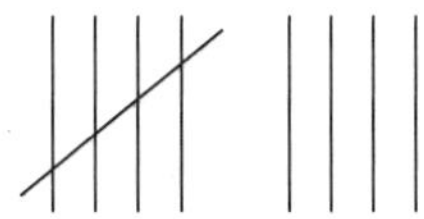

20 How many points did the Green House score?

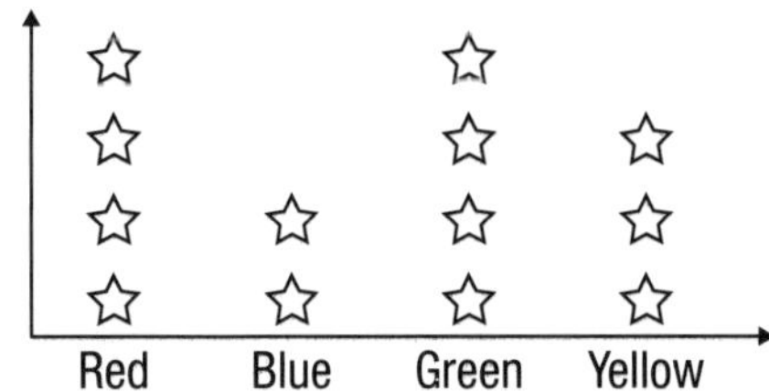

1

+	11	27	16	3	19	26
24						

2

–	27	32	16	25	44	50
13						

3

×	0	11	9	6	7	12
6						

4

×	2	11	3	8	5	1
7						

5

÷	55	35	10	50	25	40
5						

6

÷	96	56	8	40	32	88
8						

7 $400 + 900 = \square$

8 What is 85 plus 17 plus 50? $\square$

9

$$\begin{array}{r} 716 \\ -\ 435 \\ \hline \end{array}$$

10 $1900 - 180 = \square$

11 7 groups of 15 = $\square$

12 $172 \times 4 = \square$

13 Circle the numbers divisible by 5.

20 36 70 47

14 $8\overline{)96}$

15 What is the value of the underlined digit?

$1\underline{7}3\,067$ $\square$

16 Write seven hundred and twenty-four thousand, one hundred and seventeen in digits.

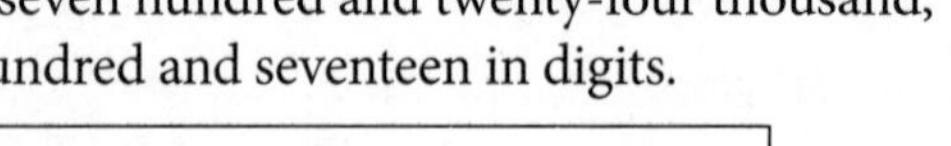

$\square$

17 What is the total cost of: $\square$

$10.25 and $11.95 ?

18 Cate bought an apple for 25c. What was Cate's change from $2.00? $\square$

19 Circle the largest value.

0.72 0.27 0.76 0.29

20 Write 3 hundredths as a decimal. $\square$

21 Round 32 096 to the nearest hundred. $\square$

22 $200 \div \square = 4$

1

$$\begin{array}{r} 246\,398 \\ +\ 427\,117 \\ \hline \end{array}$$

2 What is the difference between 6000 and 80 000?

3 There are 5 apples in a bag. How many apples are in 625 bags?

4 True or false?

887 ÷ 7 = 168

5 What fraction of the group is shaded?

6 Colour each fraction to find the answer.

$\frac{1}{10} + \frac{3}{10} =$

7

115 + 76 + 39 – 57 =

8 Draw a quarter past twelve on the clock.

9 Write the time in words.

11 : 15

10

km = 1000 m

11 Tick the most appropriate unit of measurement.

Object	mm	cm	m	km
thickness of a piece of paper				

12 What is the volume of the centicube model?

cm^3

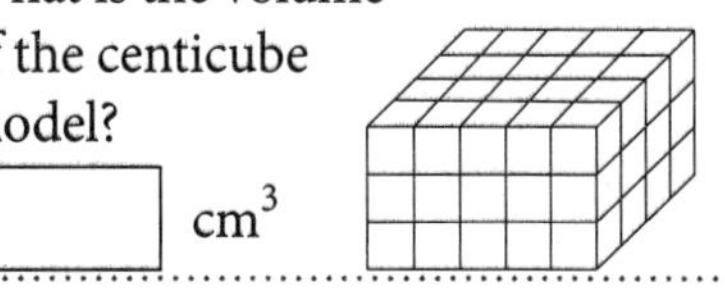

13 How many sides does this shape have?

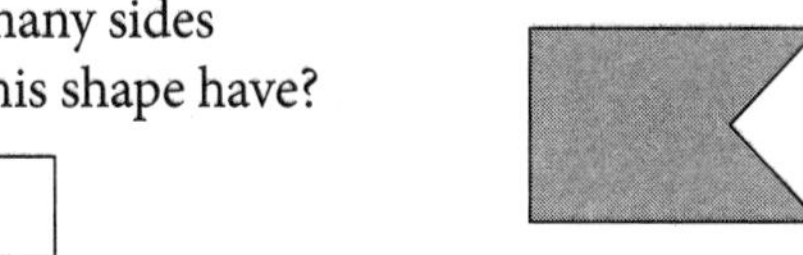

14 Name this 3D shape.

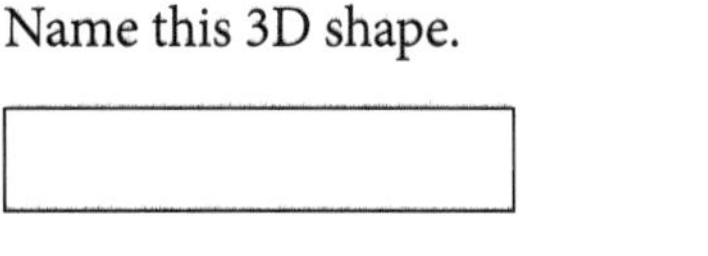

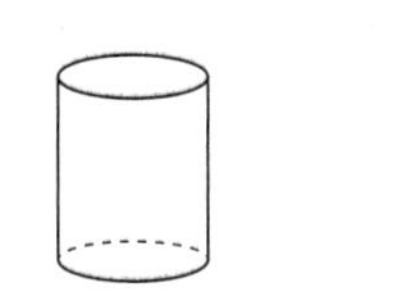

15 What is south of the pear?

16 What is at C3?

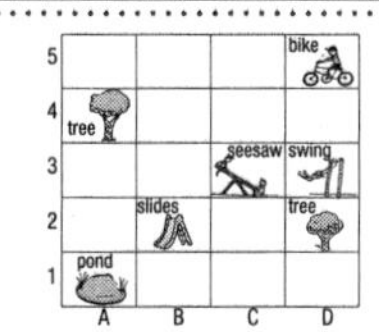

17 True or false? 100° is an obtuse angle.

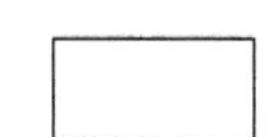

18 What is the most likely ball number to be selected from the box?

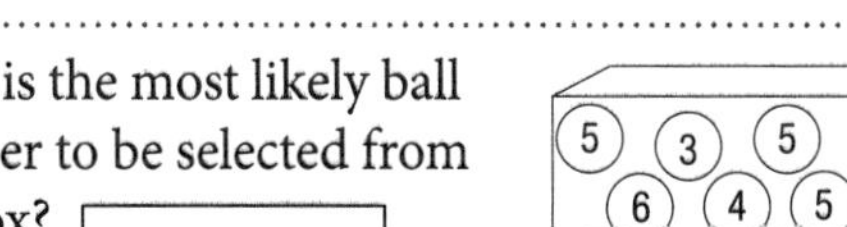

19 Rate the likelihood that you will travel to the moon this year on a scale of 0 (impossible) to 1 (certain).

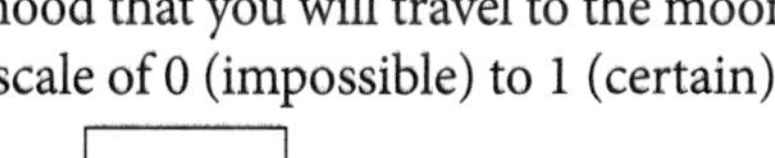

20 Complete the numbers on the tally chart.

Colour	Tally	Number
red	𝍸 𝍸 I	
green	𝍸 II	
blue	IIII	
yellow	𝍸 𝍸	

21 What was the temperature at 7 o'clock?

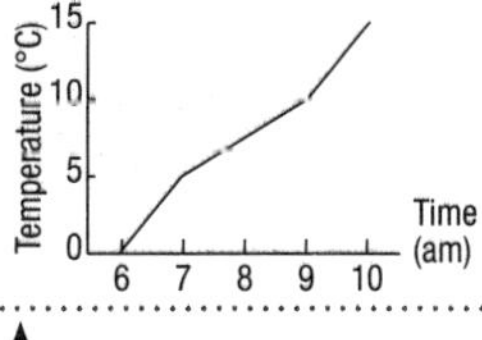

22 Draw a column graph showing:
8 pencils
6 sharpeners
11 highlighters

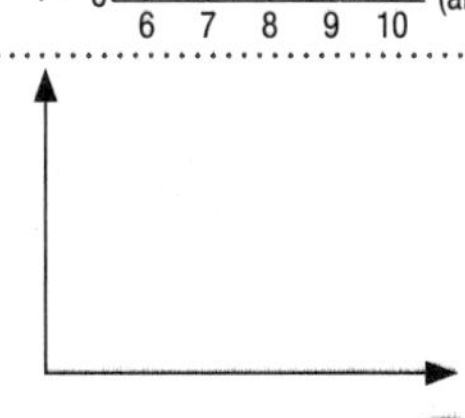

UNIT **9A**

1

+	41	20	27	18	13	36
9						

2

–	18	25	41	22	36	50
16						

3

×	4	6	9	3	7	0
6						

4

×	2	8	5	10	12	1
7						

5

÷	50	5	40	20	35	45
5						

6

÷	96	32	72	56	24	8
8						

7 $500 + 600 = \square$

8 What is 26 added to 75 and 98?

9

$$\begin{array}{r} 867 \\ -\ 491 \\ \hline \end{array}$$

10 $1400 - 170 = \square$

11 6 groups of 17 = $\square$

12 $321 \times 4 = \square$

13 Circle the numbers divisible by 9.

72 85 45 53

14 $3\overline{)87}$

15 What is the value of the underlined digit?

$3\underline{1}6\,427$

16 Write one hundred and four thousand in digits.

17 What is the total cost of:

$7.62 and $2.97?

18 Jack bought a banana for 15c.
What was Jack's change from $2.00?

19 Circle the largest value.

0.19 0.20 0.16 0.11

20 Write 20 hundredths as a decimal.

21 Round 14 001 to the nearest hundred.

22 $121 = 200 - \square$

1

$$\begin{array}{r} 798\,106 \\ +\ 432\,119 \\ \hline \end{array}$$

2 What is the difference between 300 000 and 40 000?

3 There are 8 bread rolls in a bag. How many rolls are in 193 bags?

4 True or false?

896 ÷ 8 = 112

5 What fraction of the group is shaded?

6 Colour each fraction to find the answer.

$\frac{1}{8} + \frac{5}{8} =$

7

$186 + 37 + 56 - 89 =$

8 Draw a quarter to 10 on the clock.

9 Write the time in words.

2 : 00

10

cm = 10 mm

11 Tick the most appropriate unit of measurement.

Object	mm	cm	m	km
length of the classroom				

12 What is the volume of the centicube model?

cm^3

= 1 cm^3

13 How many sides does this shape have?

14 Name this 3D shape.

15 What is west of the pineapple?

16 What is at D2?

5, 4, 3, 2, 1; A, B, C, D; bike, tree, seesaw, swing, slides, tree, pond

17 True or false? 90° is a straight angle.

18 Which fruit is least likely to be selected from the basket?

19 Rate the likelihood that you will eat some vegetables today on a scale of 0 (impossible) to 1 (certain).

20 Complete the numbers on the tally chart.

Time of day	Tally	Number
morning	卌 卌 |	
afternoon	卌 卌 卌 |	
evening	卌 ||||	

21 In what month was $3000 earnt?

$4000, $3000, $2000, $1000, 0; Jan., Feb., March, April; Month

22 Draw a column graph showing:
12 flowers
6 trees
10 plants

UNIT 10A

1

+	9	23	18	17	29	30
17						

2

–	23	30	46	47	39	50
21						

3

×	9	5	10	1	4	7
6						

4

×	0	3	8	6	10	2
7						

5

÷	10	40	60	25	55	5
5						

6

÷	24	8	64	72	40	96
8						

7 $70 + 30 = \square$

8 What is the sum of 39, 47 and 53?

9

$$\begin{array}{r} 467 \\ -\ 183 \\ \hline \end{array}$$

10 $5000 - 110 = \square$

11 8 groups of 19 = $\square$

12 $739 \times 3 = \square$

13 Circle the numbers divisible by 6.

32 48 54 69

14 $7\overline{)98}$

15 What is the value of the underlined digit?

<u>7</u>83 291

16 Write seven hundred and twenty-four thousand and one in digits.

17 What is the total cost of:

and

?

18 Lana bought a comb for 75c. What was Lana's change from $2.00?

19 Circle the largest value.

0.09 0.16 0.03 0.11

20 Write 9 hundredths as a decimal.

21 Round 17 389 to the nearest hundred.

22 $360 = \square \times 6$

UNIT 10B

1

$$\begin{array}{r} 398\,541 \\ +\ 642\,056 \\ \hline \end{array}$$

2 What is the difference between 60 000 and 400 000?

3 There are 8 carrots in a bag. How many carrots are in 246 bags?

4 True or false?

216 ÷ 3 = 72

5 What fraction of the group is shaded?

6 Colour each fraction to find the answer.

$\frac{1}{6} + \frac{3}{6} =$

7

90 + 76 + 35 – 66 =

8 Draw half past eight on the clock.

9 Write the time in words.

6 : 30

10

100 cm = ☐ m

11 Tick the most appropriate unit of measurement.

Object	mm	cm	m	km
length of a pencil case				

12 What is the volume of the centicube model?

☐ cm^3

= 1 cm^3

13 How many sides does this shape have?

14 Name this 3D shape.

15 What is north of the cherries?

16 What is at A1?

17 True or false? 330° is a full revolution.

18 What colour ball is the most likely to be selected from the box?

19 Rate the likelihood that there is art this week at school on a scale of 0 (impossible) to 1 (certain).

20 Complete the numbers on the tally chart.

Vegetable	Tally	Number
potato	𝍸 𝍸 𝍸	
pumpkin	𝍸 |||	
broccoli	𝍸		
carrot	||	

21 What was the temperature of the water at 2 minutes?

22 Draw a column graph showing:
10 squares
13 triangles
11 circles

UNIT 11A

1

+	29	52	73	46	35	60
29						

2

–	70	45	83	97	29	46
27						

3

×	7	3	8	5	4	1
9						

4

×	6	9	10	0	12	2
3						

5

÷	54	6	24	36	18	60
6						

6

÷	108	18	36	99	45	90
9						

7 Alex has 144 cards and collects another 27 cards. How many cards does Alex have altogether?

8

$$\begin{array}{r} 1367 \\ +\ 2491 \\ \hline \end{array}$$

9

$500 - 369 = \square$

10 What is the difference between 1763 and 3427?

11

$$\begin{array}{r} 486 \\ \times\ \ \ 6 \\ \hline \end{array}$$

12 Find the total cost of 16 books at $9 each.

13

$8\overline{)816}$

14 What is 400 ÷ 10?

15

$3000 + 400 + 60 + 1 = \square$

16 Show 3206 on the abacus.

Th. H T U

17 What is the change from $5 if I spend $3.25?

18 Round $23.90 to the nearest $5.00.

19 Write 0.76 on the number line.

0.6 0.7 0.8

20 Write four thousandths on the place value chart.

Units	.	Tenths	Hundredths	Thousandths

21 Round 32 106 to the nearest thousand.

22

$$\begin{array}{r} 2\ \ 4\ \ 6 \\ +\ \square\ \ 3\ \ \square \\ \hline 9\ \ \square\ \ 8 \\ \hline \end{array}$$

1 What is the total cost?

$24 116 (car) $3695 (motorbike)

2 $300\,000 - 126\,395 =$ ☐

3

$$\begin{array}{r} 4369 \\ \times \quad 5 \\ \hline \end{array}$$

4 $750 \div 6 =$ ☐

5 $\frac{1}{6} + \frac{2}{6} =$ ☐

6 Use the array to find $\frac{1}{2}$ of 20.

7 $6 \times 4 \div 3 =$ ☐

8 Write the time that is one hour later than 11:55 am.

9 John starts football training at 4:30. How long is training if it finishes at 6:00?

10 What is the most suitable unit (g or kg) for finding the mass of a box of fruit?

11 Find the perimeter of the shape.

4 m, 13 m

P = ☐ m

12 Find the total capacity of 45 mL and 125 mL.

13 How many vertices does a rectangle have?

14 How many edges does this shape have?

15 Who is sitting in front of James?

16 List the coordinates of the vertices of the triangle.

17 Circle the largest angle.

92° right angle 85°

18 What is the chance of landing on green?

19 Which number card is most likely to be selected?

20 How many apples were in the boxes?

Fruit	Tally
orange	𝍸 𝍸 𝍸
apple	𝍸 𝍸 |
pear	𝍸 𝍸

21 Draw a column graph for the data.

Shape	Number
triangle	5
circle	8
square	10

22 What was the number of gold stars awarded?

1

+	77	65	27	20	39	54
33						

2

–	70	54	73	84	96	67
24						

3

×	1	7	9	3	8	6
9						

4

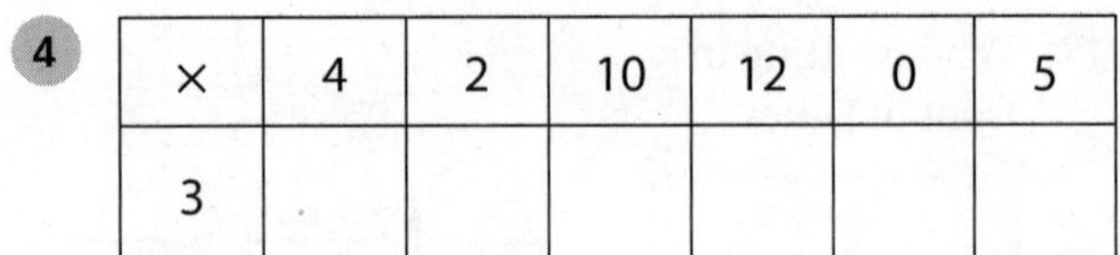

×	4	2	10	12	0	5
3						

5

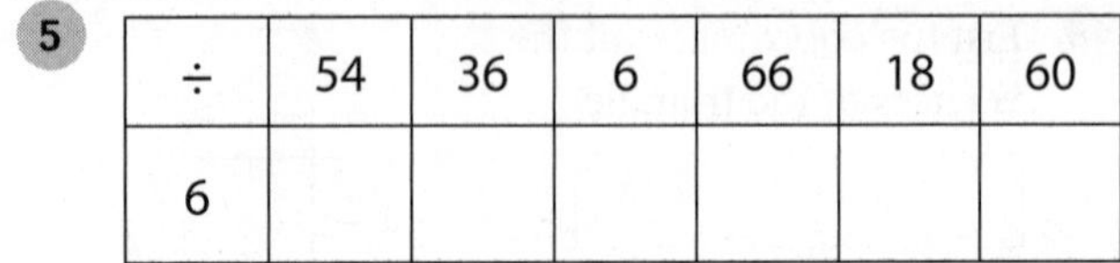

÷	54	36	6	66	18	60
6						

6

÷	9	45	108	63	36	99
9						

7 Jo has 256 nails and buys another 38 nails. How many nails does Jo have altogether?

☐

8

$$\begin{array}{r} 3427 \\ +\ 2654 \\ \hline \end{array}$$

9

$600 - 247 = \square$

10 What is 1832 take away 946?

☐

11

$$\begin{array}{r} 921 \\ \times\ \ 4 \\ \hline \end{array}$$

12 Find the total cost of 50 hats at $5 each.

☐

13

$7\overline{)217}$

14 What is 3000 ÷ 10?

☐

15

$9000 + 300 + 5 = \square$

16 Show 4700 on the abacus.

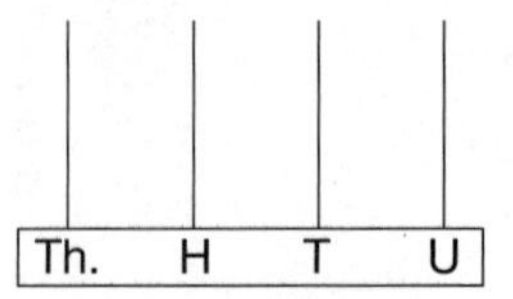

17 What is the change from $5 if I spend $3.05?

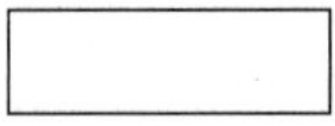

18 Round $38.29 to the nearest $5.00.

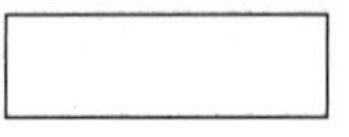

19 Write 0.45 on the number line.

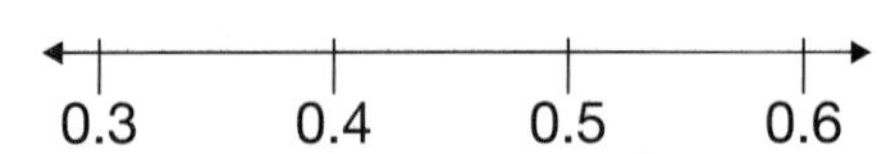

20 Write fourteen hundredths on the place value chart.

Units	.	Tenths	Hundredths	Thousandths

21 Round 113 296 to the nearest thousand.

☐

22

$$\begin{array}{r} 8\ \ 4\ \ 3 \\ -\ \square\ \ 7\ \ \square \\ \hline 2\ \ \square\ \ 8 \\ \hline \end{array}$$

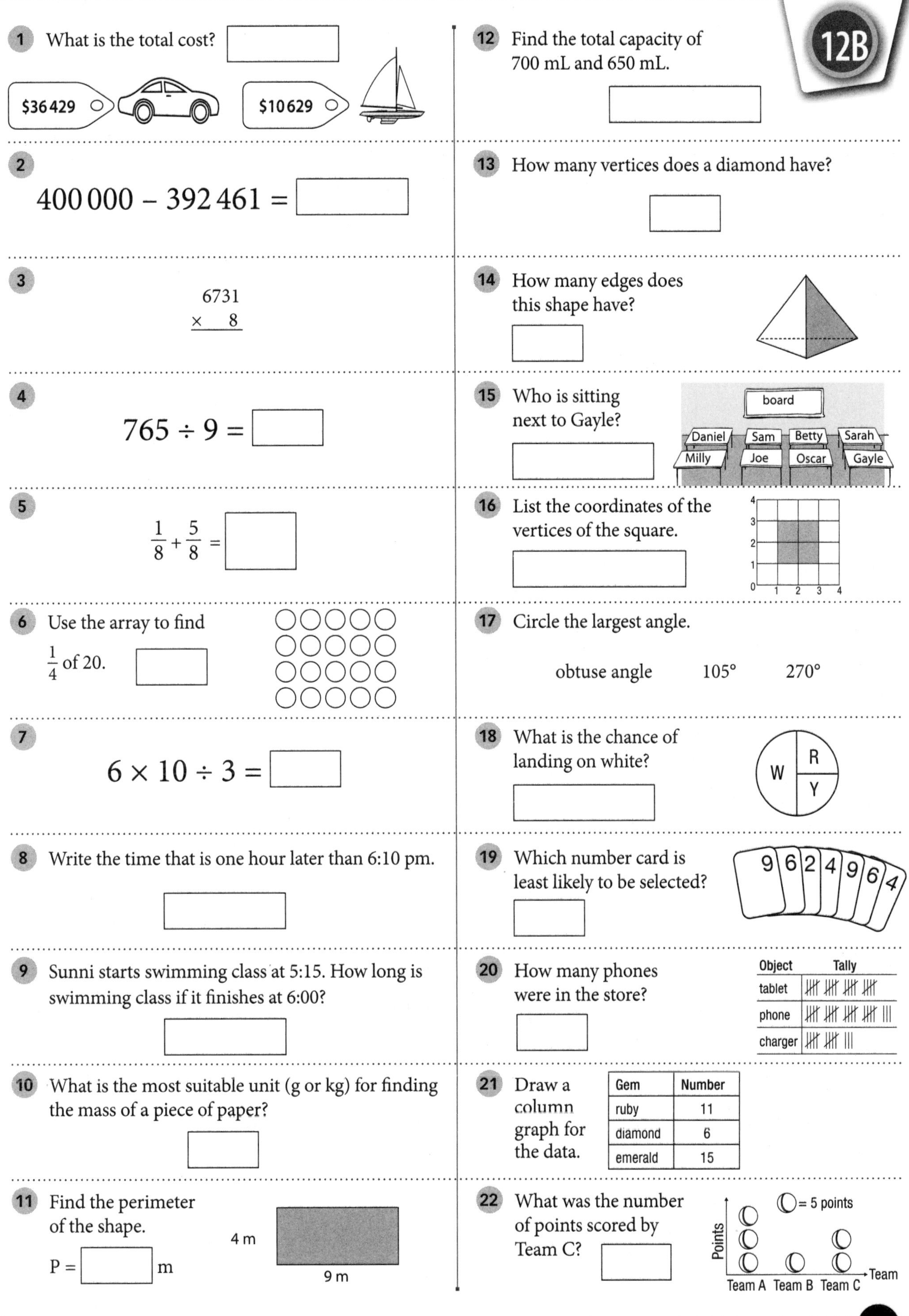

1 What is the total cost?

$36 429 $10 629

2 $400\,000 - 392\,461 =$

3
$$\begin{array}{r} 6731 \\ \times \quad 8 \\ \hline \end{array}$$

4 $765 \div 9 =$

5 $\frac{1}{8} + \frac{5}{8} =$

6 Use the array to find $\frac{1}{4}$ of 20.

7 $6 \times 10 \div 3 =$

8 Write the time that is one hour later than 6:10 pm.

9 Sunni starts swimming class at 5:15. How long is swimming class if it finishes at 6:00?

10 What is the most suitable unit (g or kg) for finding the mass of a piece of paper?

11 Find the perimeter of the shape.

P = ☐ m

12 Find the total capacity of 700 mL and 650 mL.

13 How many vertices does a diamond have?

14 How many edges does this shape have?

15 Who is sitting next to Gayle?

16 List the coordinates of the vertices of the square.

17 Circle the largest angle.

obtuse angle 105° 270°

18 What is the chance of landing on white?

19 Which number card is least likely to be selected?

20 How many phones were in the store?

Object	Tally
tablet	卌 卌 卌 卌
phone	卌 卌 卌 卌 |||
charger	卌 卌 |||

21 Draw a column graph for the data.

Gem	Number
ruby	11
diamond	6
emerald	15

22 What was the number of points scored by Team C?

1

+	17	35	29	10	42	26
54						

2

–	89	77	52	70	60	96
34						

3

×	1	4	6	8	10	7
9						

4

×	10	3	2	6	5	9
3						

5

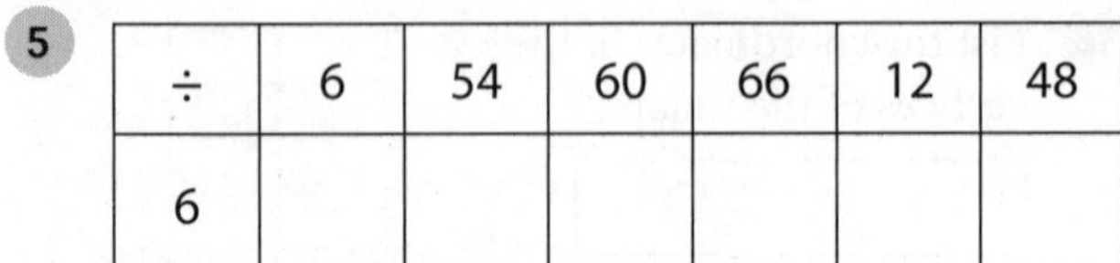

÷	6	54	60	66	12	48
6						

6

÷	27	72	45	81	90	54
9						

7 Sophie has 377 counters and she selects another 126 counters. How many counters does Sophie have altogether?

8

$$\begin{array}{r} 7721 \\ +\ 4385 \\ \hline \end{array}$$

9

$$400 - 196 = \square$$

10 What is 7436 subtract 2295?

11

$$\begin{array}{r} 365 \\ \times\ \ 5 \\ \hline \end{array}$$

12 Find the total cost of 20 lunches at $6 each.

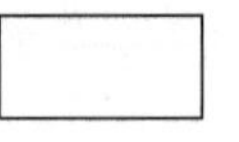

13

$$6\overline{)186}$$

14 What is 1700 ÷ 10?

15

$$7000 + 60 = \square$$

16 Show 7093 on the abacus.

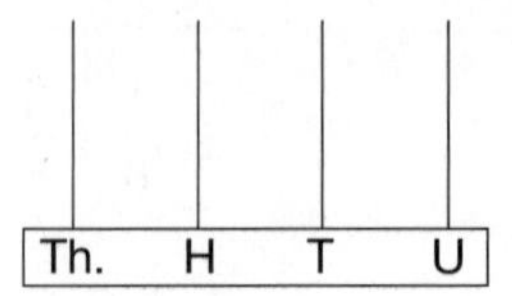

17 What is the change from $5 if I spend $4.10?

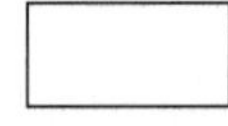

18 Round $18.55 to the nearest $5.00.

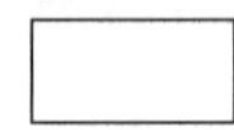

19 Write 0.15 on the number line.

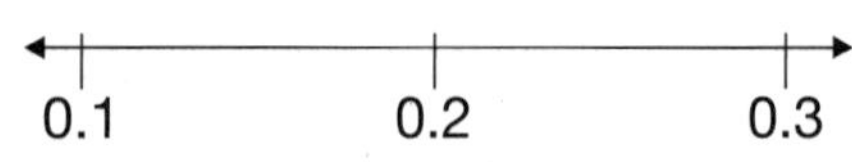

20 Write three-hundredths on the place value chart.

Units	.	Tenths	Hundredths	Thousandths

21 Round 206 985 to the nearest thousand.

22

$$\begin{array}{r} 2\ \square\ 8 \\ +\ \square\ 7\ 6 \\ \hline 4\ 9\ \square \\ \hline \end{array}$$

1 What is the total cost?

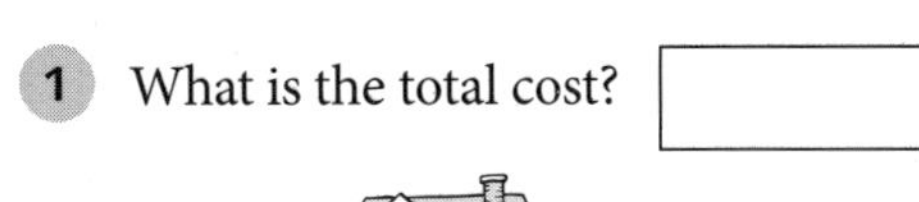
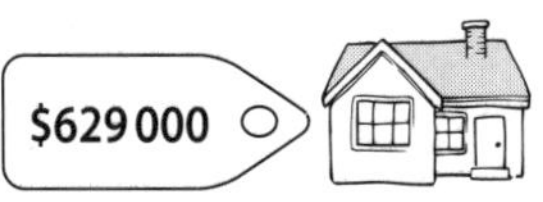

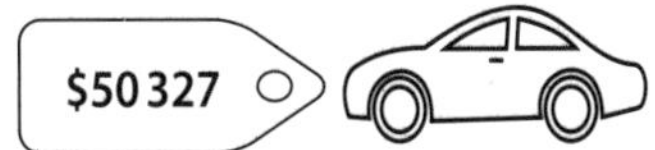

2 $600\,000 - 429\,326 =$ ☐

3

$$\begin{array}{r} 2796 \\ \times \quad 4 \\ \hline \end{array}$$

4 $552 \div 3 =$ ☐

5 $\frac{1}{10} + \frac{8}{10} =$ ☐

6 Use the array to find $\frac{1}{5}$ of 20. ☐

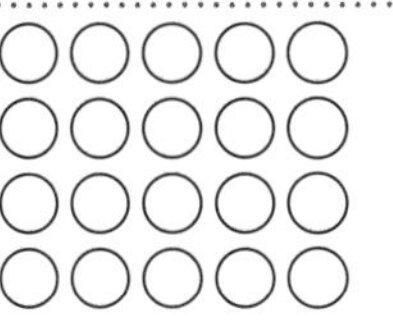

7 $12 \times 8 \div 4 =$ ☐

8 Write the time that is one hour later than 2:05 pm.

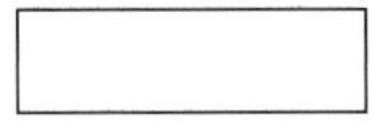

9 Zara starts art class at 4:20. How long is art class if it finishes at 5:15?

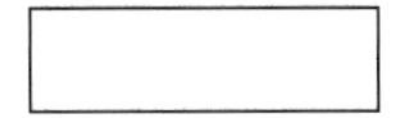

10 What is the most suitable unit (g or kg) for finding the mass of a large bag of potatoes?

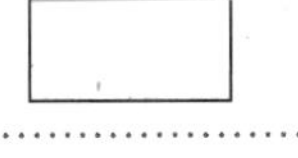

11 Find the perimeter of the shape.

P = ☐ m

12 Find the total capacity of 350 mL and 900 mL.

13 How many vertices does an octagon have?

14 How many edges does this shape have?

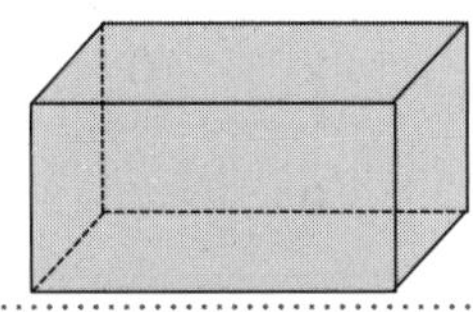

15 Who is sitting behind Lily?

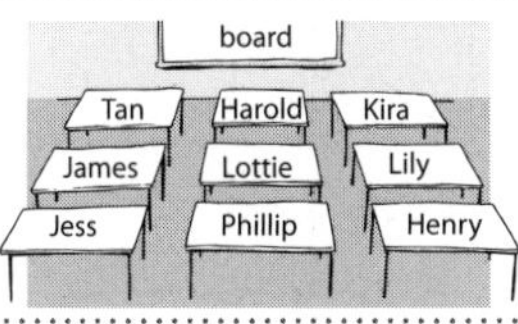

16 List the coordinates of the vertices of the shape.

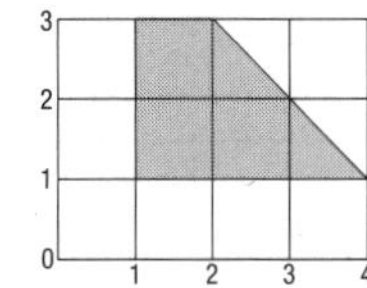

17 Circle the largest angle.

45° acute angle right angle

18 What is the chance of landing on blue?

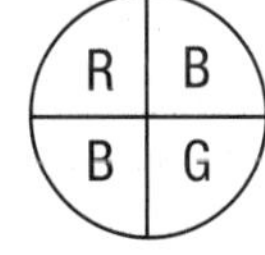

19 Which number card has a 50% chance of being selected?

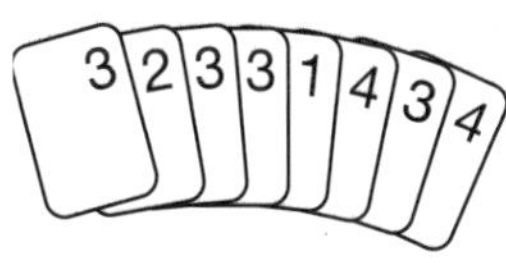

20 How many cars were parked on Friday?

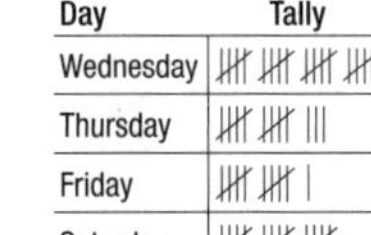

Day	Tally
Wednesday	卌 卌 卌 卌
Thursday	卌 卌 |||
Friday	卌 卌 |
Saturday	卌 卌 卌

21 Draw a column graph for the data.

Location	Number
pool	5
lake	8
ocean	10

22 What was the number of rap songs selected?

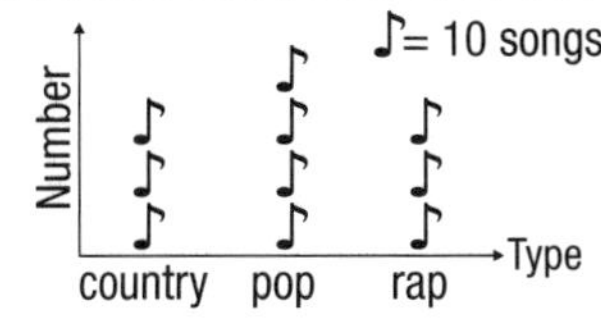

1

+	36	64	72	18	85	59
18						

2

–	54	44	79	60	96	85
36						

3

×	10	2	4	9	8	6
9						

4

×	1	9	7	11	3	5
3						

5

÷	60	24	54	30	72	18
6						

6

÷	9	99	36	63	18	90
9						

7 Colin has 225 flowers to sell and he picks another 117 flowers. How many flowers does Colin have altogether to sell?

8

$$\begin{array}{r} 6256 \\ +\ 3309 \\ \hline \end{array}$$

9 $500 - 177 = \square$

10 What is 1325 less than 3720?

11

$$\begin{array}{r} 117 \\ \times\ \ 8 \\ \hline \end{array}$$

12 Find the total cost of 19 drinks at $3.00 each.

13

$$9\overline{)279}$$

14 What is 2300 ÷ 10?

15 $6000 + 900 + 80 + 5 = \square$

16 Show 1130 on the abacus.

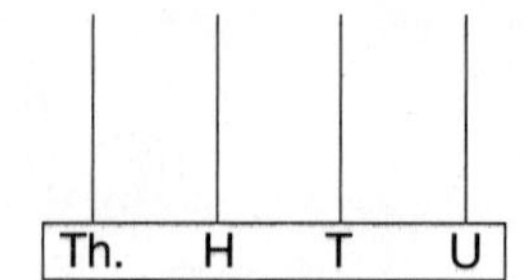

17 What is the change from $5 if I spend $2.55?

18 Round $32.05 to the nearest $5.00.

19 Write 0.81 on the number line.

0.7 0.8 0.9

20 Write thirteen-hundredths on the place value chart.

Units	.	Tenths	Hundredths	Thousandths

21 Round 8976 to the nearest thousand.

22

$$\begin{array}{r} \square\ 9\ 8 \\ -\ 5\ \square\ 9 \\ \hline 2\ 6\ \square \\ \hline \end{array}$$

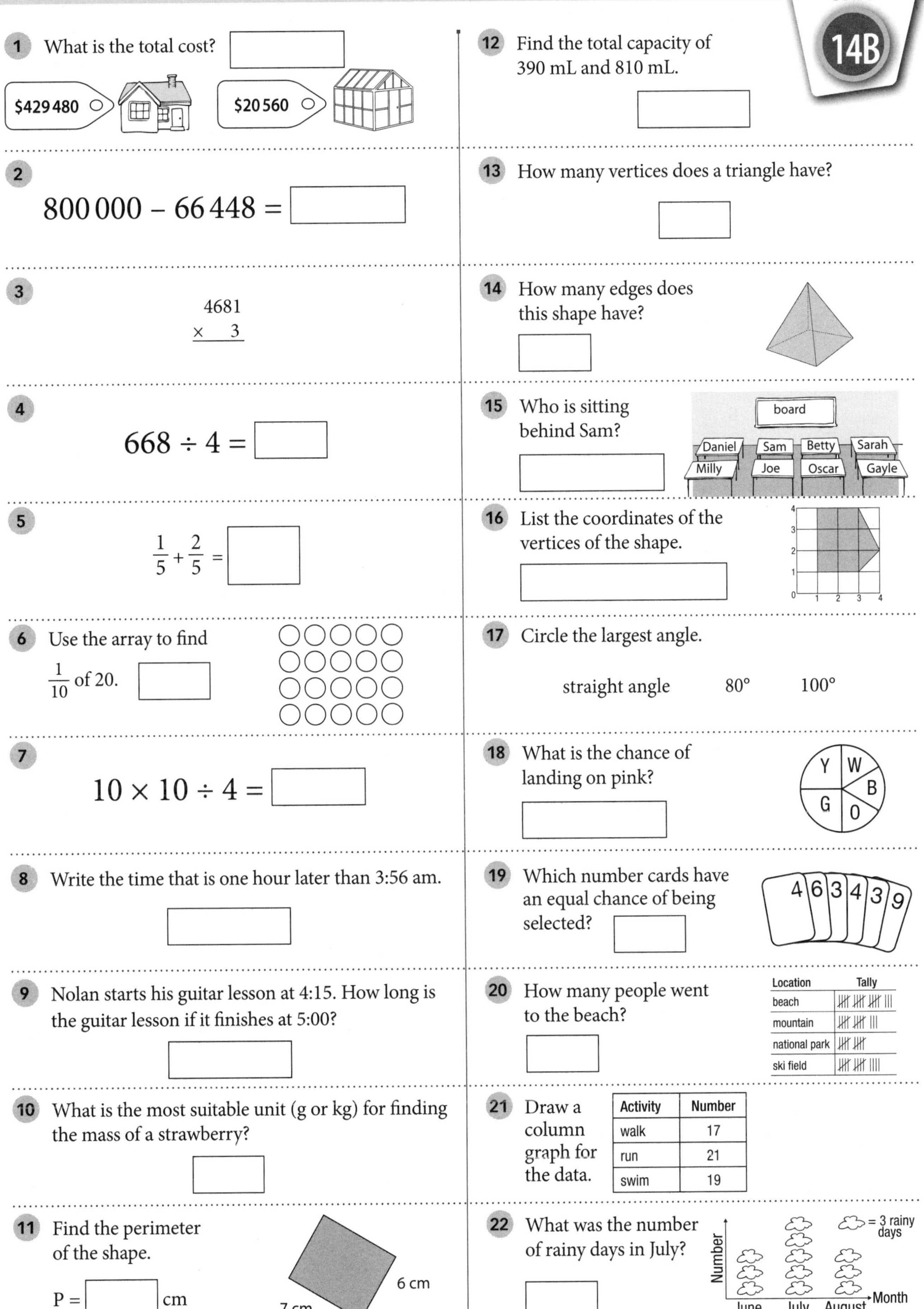

1 What is the total cost? ☐

2 $800\,000 - 66\,448 =$ ☐

3
$$\begin{array}{r} 4681 \\ \times \quad 3 \\ \hline \end{array}$$

4 $668 \div 4 =$ ☐

5 $\frac{1}{5} + \frac{2}{5} =$ ☐

6 Use the array to find $\frac{1}{10}$ of 20. ☐

7 $10 \times 10 \div 4 =$ ☐

8 Write the time that is one hour later than 3:56 am. ☐

9 Nolan starts his guitar lesson at 4:15. How long is the guitar lesson if it finishes at 5:00? ☐

10 What is the most suitable unit (g or kg) for finding the mass of a strawberry? ☐

11 Find the perimeter of the shape.

P = ☐ cm

12 Find the total capacity of 390 mL and 810 mL. ☐

13 How many vertices does a triangle have? ☐

14 How many edges does this shape have? ☐

15 Who is sitting behind Sam? ☐

16 List the coordinates of the vertices of the shape. ☐

17 Circle the largest angle.

straight angle   80°   100°

18 What is the chance of landing on pink? ☐

19 Which number cards have an equal chance of being selected? ☐

20 How many people went to the beach? ☐

Location	Tally
beach	卌 卌 卌 |||
mountain	卌 卌 |||
national park	卌 卌
ski field	卌 卌 ||||

21 Draw a column graph for the data.

Activity	Number
walk	17
run	21
swim	19

22 What was the number of rainy days in July? ☐

1

+	67	37	46	18	29	50
27						

2

–	49	64	83	50	72	97
18						

3

×	11	2	12	7	5	3
9						

4

×	10	8	1	4	9	6
3						

5

÷	36	24	30	54	12	60
6						

6

÷	54	90	27	18	72	81
9						

7 Simon the baker has 152 cupcakes and he bakes another 139 cupcakes. How many cupcakes does Simon have altogether?

8

$$\begin{array}{r} 3177 \\ +\ 4163 \\ \hline \end{array}$$

9

$600 - 288 = \square$

10 What is 9833 take away 3177?

11

$$\begin{array}{r} 408 \\ \times\ \ 7 \\ \hline \end{array}$$

12 Find the total cost of 17 games at $8 each.

13

$4\overline{)248}$

14 What is 8000 ÷ 10?

15

$20\,000 + 8000 + 700 + 20 + 9 = \square$

16 Show 9305 on the abacus.

Th. H T U

17 What is the change from $5 if I spend $3.80?

18 Round $55.29 to the nearest $5.00.

19 Write 0.55 on the number line.

0.4 0.5 0.6

20 Write ninety-hundredths on the place value chart.

Units	.	Tenths	Hundredths	Thousandths

21 Round 49 987 to the nearest thousand.

22

$$\begin{array}{r} 3\ \square\ 4 \\ +\ 4\ 8\ \square \\ \hline \square\ 3\ 3 \\ \hline \end{array}$$

1 What is the total cost? [ ]

$40 685 $50 209

2

$600\,000 - 326\,417 =$ [ ]

3

$$\begin{array}{r} 7291 \\ \times \quad 8 \\ \hline \end{array}$$

4

$966 \div 6 =$ [ ]

5

$\frac{1}{4} + \frac{1}{4} =$ [ ]

6 Use the array to find $\frac{1}{20}$ of 20. [ ]

7

$8 \times 12 \div 3 =$ [ ]

8 Write the time that is one hour later than 8:20 pm.

[ ]

9 Declan starts ballet class at 7:10. How long is ballet class if it finsihes at 8:30?

[ ]

10 What is the most suitable unit (g or kg) for finding the mass of a lion?

[ ]

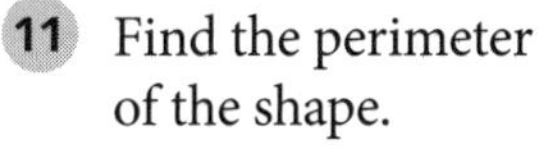

11 Find the perimeter of the shape.

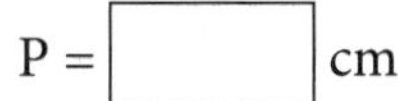

P = [ ] cm

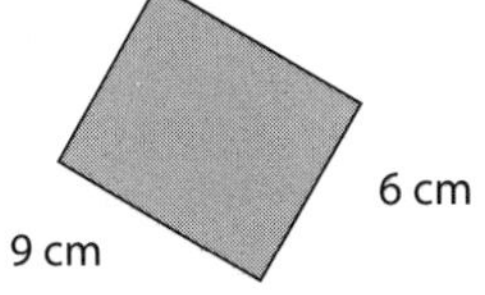

12 Find the total capacity of 400 mL and 700 mL.

[ ]

13 How many vertices does a parallelogram have?

[ ]

14 How many edges does this shape have?

[ ]

15 Who is sitting at the centre of the back of the classroom? [ ]

board
Tan Harold Kira
James Lottie Lily
Jess Phillip Henry

16 List the coordinates of the vertices of the shape.

[ ]

17 Circle the largest angle.

obtuse angle 215° full revolution

18 What is the chance of landing on grey?

[ ]

Y G B W G

19 Which number card is most likely to be selected?

[ ]

9 7 8 7 9 7

20 How many people preferred lemon?

[ ]

Ice-cream	Tally
chocolate	𝍸 𝍸 𝍸
lemon	𝍸 𝍸 ||
strawberry	𝍸 𝍸 𝍸 ||
mango	𝍸 ||

21 Draw a column graph for the data.

Direction	Number
north	22
south	17
east	10
west	15

22 What was the number of cars parked on Wednesday? [ ]

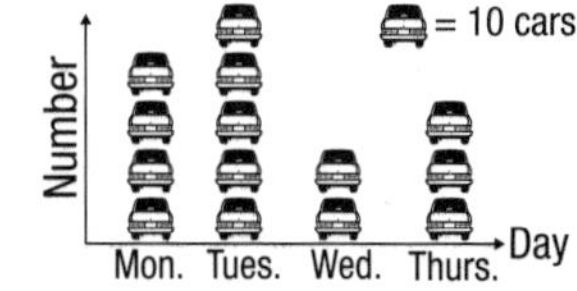

1

+	40	37	19	26	46	59
35						

2

–	49	63	85	92	21	30
20						

3

×	4	7	9	8	10	3
9						

4

×	1	6	5	11	12	2
7						

5

÷	96	8	32	16	40	80
8						

6

÷	72	54	48	24	36	66
6						

7 Caitlin has 476 songs on her phone. She merges a playlist which has another 257 songs. How many songs does she have altogether?

8

$$\begin{array}{r} 6367 \\ +\ 2178 \\ \hline \end{array}$$

9 What is the difference between 1800 and 365?

10 Is 432 – 169:

A less than 376? **B** greater than 376?
C equal to 376?

11 Bananas are $4.00 a bag. What is the total cost of 20 bags?

12

$$\begin{array}{r} 963 \\ \times\ 4 \\ \hline \end{array}$$

13 Which number is divisible by 4?

A 62 **B** 72
C 86 **D** 114

14 What is 9000 divided by 10?

15 What is the digit in the thousands place in 17 683?

A 1 **B** 7
C 6 **D** 8

16 What value is shown on the abacus?

17 Rose bought this cake. What was Rose's change from $5?

18 What is $41.65 rounded to the nearest dollar?

19 Which is the largest value?

A 0.16 **B** 0.9
C 0.07 **D** 0.115

20 What is 39 hundredths written as a decimal?

A 390 **B** 3.9
C 0.39 **D** 0.039

21 Round 32 069 to the nearest hundred.

22 What is the missing value?

90 × ☐ = 270

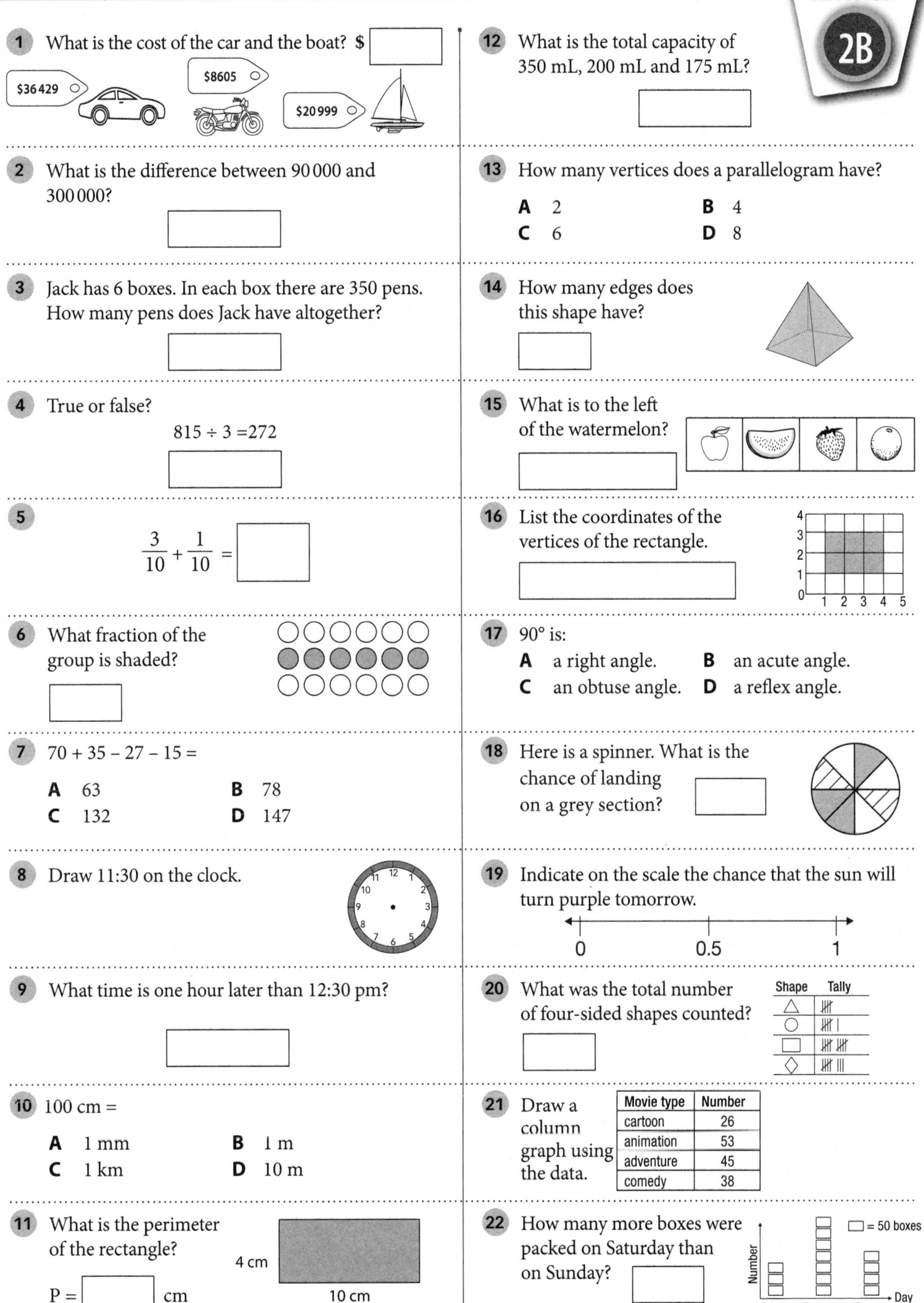

1 What is the cost of the car and the boat? $ ______

$36 429 (car) $8605 (motorbike) $20 999 (boat)

2 What is the difference between 90 000 and 300 000?

3 Jack has 6 boxes. In each box there are 350 pens. How many pens does Jack have altogether?

4 True or false?

815 ÷ 3 =272

5

$$\frac{3}{10} + \frac{1}{10} = \square$$

6 What fraction of the group is shaded?

7 70 + 35 – 27 – 15 =

A 63 **B** 78
C 132 **D** 147

8 Draw 11:30 on the clock.

9 What time is one hour later than 12:30 pm?

10 100 cm =

A 1 mm **B** 1 m
C 1 km **D** 10 m

11 What is the perimeter of the rectangle?

4 cm
10 cm

P = ______ cm

12 What is the total capacity of 350 mL, 200 mL and 175 mL?

13 How many vertices does a parallelogram have?

A 2 **B** 4
C 6 **D** 8

14 How many edges does this shape have?

15 What is to the left of the watermelon?

16 List the coordinates of the vertices of the rectangle.

17 90° is:

A a right angle. **B** an acute angle.
C an obtuse angle. **D** a reflex angle.

18 Here is a spinner. What is the chance of landing on a grey section?

19 Indicate on the scale the chance that the sun will turn purple tomorrow.

0 0.5 1

20 What was the total number of four-sided shapes counted?

Shape	Tally
△	𝍸
○	𝍸 I
□	𝍸 𝍸
◇	𝍸 III

21 Draw a column graph using the data.

Movie type	Number
cartoon	26
animation	53
adventure	45
comedy	38

22 How many more boxes were packed on Saturday than on Sunday?

□ = 50 boxes
Number
Day
Friday Saturday Sunday

1

$$3098 + 1112 = \square$$

2 1200 – 56 =

A 144
B 1144
C 1154
D 1256

3 There are 8 small bottles of water in a box. Chris buys 124 boxes for a race. How many bottles of water does Chris have altogether?

A 132
B 872
C 992
D 996

4 What is 1700 divided by 10?

A 17 000
B 170
C 17
D 1.7

5 Sam wrote number sentences on cards. Which card equals 30 613?

30 000 + 600 + 90 + 1	30 + 6 + 10 + 3	30 000 + 3 + 10 + 600	3000 + 3 + 60 + 10
A	**B**	**C**	**D**

Answers

Unit 1A page 8

1. 20, 16, 24, 27, 29, 23
2. 23, 21, 17, 10, 12, 16
3. 5, 15, 35, 30, 40, 20
4. 32, 40, 64, 72, 80, 16
5. 10, 12, 5, 7, 3, 11
6. 11, 4, 3, 2, 6, 10
7. 200, 2000
8. 1328
9. 133
10. 115
11. 234
12. 6
13. 10
14. 11
15. eighty-three thousand, four hundred and sixty-nine
16. 47 948
17. 25c
18. 10
19. 0.3
20. 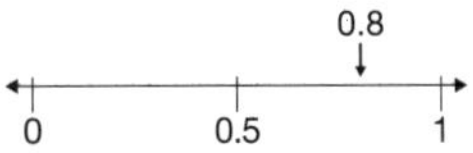

21. 680
22. 30

Unit 1B page 9

1. 230 000
2. 15 015
3. 847
4. 115
5. $\frac{1}{8}$
6. 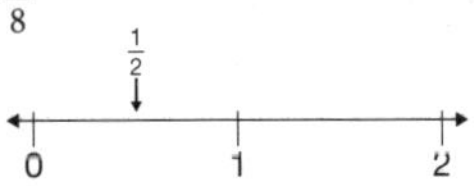

7. 48
8. half past four or four-thirty
9. 6:45
10. m
11. 26 mm, 2.6 cm
12. parent/teacher to check
13. pentagon
14. 6
15. L
16. parent/teacher to check
17. 3
18. parent/teacher to check
19. head, tail
20.

Fruit	Tally
apples	卌 卌 卌 卌 卌 ||||

21. 6
22. 2 m

Unit 2A page 10

1. 18, 30, 23, 26, 19, 20
2. 22, 16, 9, 4, 13, 20
3. 30, 45, 15, 40, 10, 50
4. 32, 40, 8, 72, 56, 0
5. 4, 3, 6, 7, 10, 1
6. 4, 10, 2, 1, 7, 8
7. 220, 2200
8. 1205
9. 439
10. 346
11. 230
12. 4
13. 10
14. 5
15. one hundred and four thousand
16. 53 090
17. 45c
18. 40
19. 0.7
20. 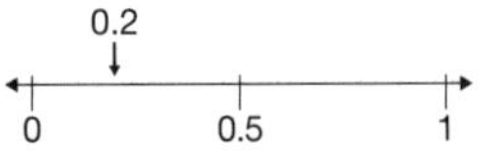

21. 420
22. 15

Unit 2B page 11

1. 470 000
2. 10 204
3. 1136
4. 116
5. $\frac{1}{4}$
6. 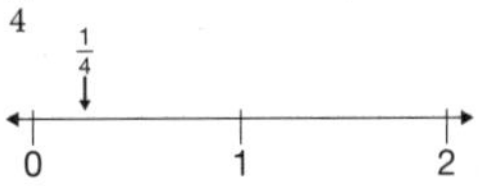

7. 72
8. eleven o'clock
9. 3:15
10. kg
11. 51/52 mm, 5.1/5.2 cm
12. parent/teacher to check
13. equilateral
14. 4
15. B
16. parent/teacher to check
17. 4
18. parent/teacher to check
19. 1, 2, 3, 4, 5, 6
20.

Animal	Tally
cats	卌 卌 卌 卌 卌 ||

21. 21
22. 1.4 m

Unit 3A page 12

1. 15, 14, 26, 29, 18, 11
2. 3, 15, 6, 8, 12, 24
3. 0, 40, 35, 15, 30, 5
4. 16, 40, 64, 80, 32, 72
5. 11, 4, 1, 5, 7, 3
6. 3, 10, 2, 4, 7, 11
7. 300, 3000
8. 611
9. 526
10. 216
11. 504
12. 6
13. 10
14. 7
15. three hundred and ten thousand, six hundred and twenty-six
16. 67 420
17. 65c
18. 10
19. 0.4
20. 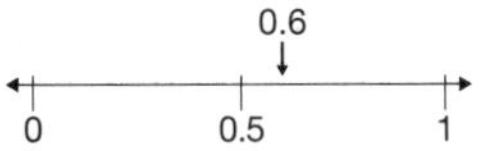

21. 770
22. 31

Unit 3B page 13

1. 1 600 000
2. 9590
3. 1043
4. 192
5. $\frac{1}{5}$
6. 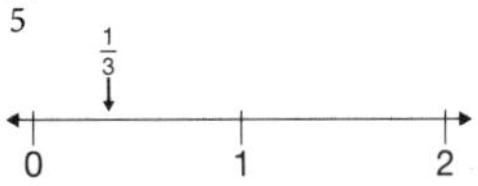

7. 66
8. one o'clock
9. 5:45
10. m
11. 36 mm, 3.6 cm
12. parent/teacher to check
13. hexagon
14. 5
15. 7
16. parent/teacher to check
17. 4
18. parent/teacher to check
19. B, G, R
20.

Insect	Tally
flies	卌 卌 卌 卌 卌 卌 卌 卌 卌 ||

21. 4
22. 10 am

Unit 4A page 14

1. 21, 24, 17, 15, 30, 23
2. 21, 9, 16, 1, 12, 18
3. 5, 15, 40, 35, 45, 30
4. 16, 40, 80, 32, 0, 56
5. 11, 4, 3, 6, 2, 5
6. 6, 10, 2, 9, 4, 5
7. 320, 3200
8. 737
9. 513
10. 536
11. 243
12. 10
13. 10
14. 3
15. four hundred and nine thousand, three hundred and one
16. 32 460
17. 70c
18. 15
19. 1.2
20. 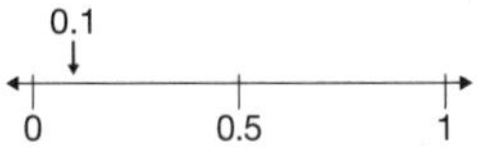

21. 900
22. 26

Unit 4B page 15

1. 750 000
2. 17 699
3. 1134
4. 120
5. $\frac{1}{8}$
6. 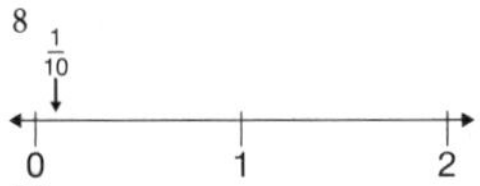

7. 50
8. half past six or six-thirty
9. 12:15
10. L
11. 44 mm, 4.4 cm
12. parent/teacher to check
13. isosceles
14. 6
15. 3
16. parent/teacher to check
17. 5
18. parent/teacher to check
19. head tail, tail head, head head, tail tail
20.

Colour	Tally
red	卌 ||||
blue	卌 卌 卌 卌 卌 ||
green	卌 |

21. golf ball
22. $2000

Unit 5A page 16

1. 21, 27, 29, 23, 20, 22
2. 16, 5, 7, 13, 10, 19
3. 15, 20, 40, 45, 35, 10
4. 8, 48, 80, 40, 56, 72
5. 7, 3, 10, 5, 8, 1
6. 1, 3, 6, 10, 8, 2
7. 860, 8600
8. 601
9. 414
10. 116
11. 301
12. 10
13. 10
14. 8
15. one hundred and seventeen thousand and thirty-nine
16. 324 204
17. 40c
18. 20
19. 2.1
20. 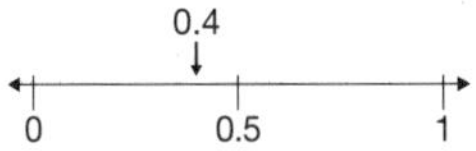

21. 310
22. 41

Unit 5B page 17

1. 808 000
2. 63 155
3. 1104
4. 112
5. $\frac{1}{3}$
6. $\frac{1}{8}$

   0 1 2
7. 132
8. half past twelve or twelve-thirty
9. 9:15
10. kg
11. 21 mm, 2.1 cm
12. parent/teacher to check
13. scalene
14. 8
15. E
16. parent/teacher to check
17. 6
18. parent/teacher to check
19. 1, 2, 3, 4, 5
20.

Drink	Tally
juice	𝍸 𝍸 𝍸 |
milk	𝍸 𝍸 𝍸 ||||
water	||||

21. 3
22. 3

Unit 6A page 18

1. 36, 40, 32, 28, 47, 29
2. 25, 12, 18, 3, 34, 10
3. 6, 18, 54, 42, 60, 48
4. 77, 42, 0, 28, 49, 84
5. 4, 12, 8, 3, 11, 5
6. 11, 3, 6, 10, 4, 9
7. 1400
8. 83
9. 224
10. 500
11. 75
12. 1304
13. 30, 15
14. 17
15. 8 hundred
16. 276 408
17. $10.89
18. 90c
19. 0.66
20. 0.36
21. 15 300
22. 10

Unit 6B page 19

1. 384 280
2. 120 000
3. 2142
4. false
5. $\frac{1}{4}$
6. $\frac{2}{4} = \frac{1}{2}$
7. 37
8. 
9. half past nine or nine-thirty
10. 1000
11. cm
12. 20
13. 4
14. rectangular prism
15. banana
16. playground
17. false
18. 3
19. parent/teacher to check
20. Monday 12, Tuesday 15, Wednesday 9
21. 6 seconds
22. parent/teacher to check

Unit 7A page 20

1. 39, 36, 30, 34, 42, 25
2. 10, 7, 19, 25, 18, 36
3. 6, 36, 24, 18, 66, 60
4. 77, 7, 63, 84, 28, 21
5. 6, 10, 2, 3, 9, 1
6. 1, 8, 3, 10, 6, 2
7. 900
8. 163
9. 119
10. 1160
11. 117
12. 1482
13. 36, 44
14. 13
15. 3 tens
16. 308 694
17. $16.48
18. $1.15
19. 0.96
20. 0.72
21. 72 400
22. 98

Unit 7B page 21

1. 1 279 845
2. 830 000
3. 1944
4. false
5. $\frac{1}{5}$
6. $\frac{2}{5}$
7. 152
8. 
9. half past one or one-thirty
10. 100
11. km
12. 6
13. 5
14. square-based pyramid
15. cherries
16. bike shed
17. true
18. G
19. parent/teacher to check
20. September 8, October 5, November 12, December 11
21. 10 years old
22. parent/teacher to check

Revision 1A page 22

1. 32, 30, 18, 26, 21, 49
2. 11, 2, 4, 17, 33, 26
3. 32, 48, 24, 64, 72, 8
4. 10, 35, 50, 55, 60, 25
5. 12, 7, 8, 11, 4, 3
6. 6, 10, 12, 4, 9, 3
7. 1200, 2400
8. 126
9. 124
10. 300, 500
11. 80
12. 2
13. 24, 48
14. 7
15. ninety-six thousand and forty
16. 81 005
17. $1.15
18. C
19. 0.9
20. C
21. 4800, 27 600, 72 100
22. 68

Revision 1B page 23

1. 712 815
2. 252 366
3. 546
4. false
5. 3
6. $\frac{7}{8}$
7. 129
8. 12:45 or a quarter to 1
9. D
10. m, cm
11. B
12. 18
13. 6
14. 4
15. pear
16. 5W
17. false, true, false
18. $\frac{2}{6} = \frac{1}{3}$
19. G, B, R
20. 8
21. 6
22. 20 cm

NAPLAN-style Test 1 page 24

1. B
2. A
3. C
4. 9
5. D
6. C
7. C
8. A
9. B
10. B
11. A
12. D
13. B
14. 5
15. D
16. star
17. B
18. B
19. C
20. 4

Unit 8A page 28

1. 35, 51, 40, 27, 43, 50
2. 14, 19, 3, 12, 31, 37
3. 0, 66, 54, 36, 42, 72
4. 14, 77, 21, 56, 35, 7
5. 11, 7, 2, 10, 5, 8
6. 12, 7, 1, 5, 4, 11
7. 1300
8. 152
9. 281
10. 1720
11. 105
12. 688
13. 20, 70
14. 12
15. 7 tens of thousands
16. 724 117
17. $22.20
18. $1.75
19. 0.76
20. 0.03
21. 32 100
22. 50

Unit 8B page 29

1. 673 515
2. 74 000
3. 3125
4. false
5. $\frac{1}{6}$
6. $\frac{4}{10} = \frac{2}{5}$
7. 173
8.
9. a quarter past eleven or eleven-fifteen
10. 1
11. mm
12. 60
13. 5
14. cylinder
15. grapes
16. see-saw
17. true
18. 5
19. parent/teacher to check
20. red 11, green 7, blue 4, yellow 10
21. 5 °C
22. parent/teacher to check

Unit 9A page 30

1. 50, 29, 36, 27, 22, 45
2. 2, 9, 25, 6, 20, 34
3. 24, 36, 54, 18, 42, 0
4. 14, 56, 35, 70, 84, 7
5. 10, 1, 8, 4, 7, 9
6. 12, 4, 9, 7, 3, 1
7. 1100
8. 199
9. 376
10. 1230
11. 102
12. 1284
13. 72, 45
14. 29
15. one ten-thousand
16. 104 000
17. \$10.59
18. \$1.85
19. 0.20
20. 0.20
21. 14 000
22. 79

Unit 9B page 31

1. 1 230 225
2. 260 000
3. 1544
4. true
5. $\frac{1}{3}$
6. $\frac{6}{8} = \frac{3}{4}$
7. 190
8. 
9. two o'clock
10. 1
11. m
12. 8
13. 6
14. cone
15. pear
16. tree
17. false
18. orange
19. parent/teacher to check
20. morning 11, afternoon 16, evening 9
21. February
22. parent/teacher to check

Unit 10A page 32

1. 26, 40, 35, 34, 46, 47
2. 2, 9, 25, 26, 18, 29
3. 54, 30, 60, 6, 24, 42
4. 0, 21, 56, 42, 70, 14
5. 2, 8, 12, 5, 11, 1
6. 3, 1, 8, 9, 5, 12
7. 100
8. 139
9. 284
10. 4890
11. 152
12. 2217
13. 48, 54
14. 14
15. 7 hundred thousand
16. 724 001
17. \$19.35
18. \$1.25
19. 0.16
20. 0.09
21. 17 400
22. 60

Unit 10B page 33

1. 1 040 597
2. 340 000
3. 1968
4. true
5. $\frac{1}{8}$
6. $\frac{4}{6} = \frac{2}{3}$
7. 135
8. (clock)
9. half past six or six-thirty
10. 1
11. cm
12. 50
13. 6
14. triangular prism
15. orange
16. pond
17. false
18. P
19. parent/teacher to check
20. potato 15, pumpkin 8, broccoli 6, carrot 2
21. 60 °C
22. parent/teacher to check

Unit 11A page 34

1. 58, 81, 102, 75, 64, 89
2. 43, 18, 56, 70, 2, 19
3. 63, 27, 72, 45, 36, 9
4. 18, 27, 30, 0, 36, 6
5. 9, 1, 4, 6, 3, 10
6. 12, 2, 4, 11, 5, 10
7. 171
8. 3858
9. 131
10. 1664
11. 2916
12. \$144
13. 102
14. 40
15. 3461
16. Th. H T U
17. \$1.75
18. \$25
19. 0.6 0.7 0.8 (arrow at 0.76)
20. 0.004
21. 32 000
22. 246 + 732 = 978

Unit 11B page 35

1. \$27 811
2. 173 605
3. 21 845
4. 125
5. $\frac{3}{6} = \frac{1}{2}$
6. 10
7. 8
8. 12:55 pm
9. 1 hour 30 minutes
10. kg
11. 34
12. 170 mL
13. 4
14. 9
15. Tan
16. (1, 1) (4, 1) (1, 3)
17. 92°
18. 25% or $\frac{1}{4}$
19. 4
20. 11
21. parent/teacher to check
22. 20

Unit 12A page 36

1. 110, 98, 60, 53, 72, 87
2. 46, 30, 49, 60, 72, 43
3. 9, 63, 81, 27, 72, 54
4. 12, 6, 30, 36, 0, 15
5. 9, 6, 1, 11, 3, 10
6. 1, 5, 12, 7, 4, 11
7. 294
8. 6081
9. 353
10. 886
11. 3684
12. \$250
13. 31
14. 300
15. 9305
16. 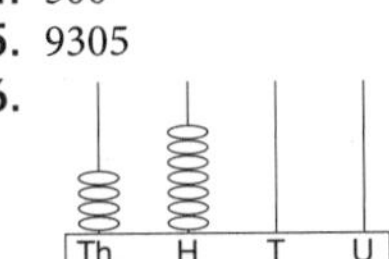

17. \$1.95
18. \$40
19. 0.3 0.4 0.5 (arrow at 0.45)
20. 0.14 or 0.140
21. 113 000
22. 843 − 575 = 268

Unit 12B page 37

1. \$47 058
2. 7539
3. 53 848
4. 85
5. $\frac{6}{8} = \frac{3}{4}$
6. 5
7. 20
8. 7:10 pm
9. 45 minutes
10. g
11. 26
12. 1350 mL or 1.35 L
13. 4
14. 6
15. Oscar
16. (1, 1) (3, 1) (3, 3) (1, 3)
17. 270°
18. 50% or $\frac{1}{2}$
19. 2
20. 23
21. parent/teacher to check
22. 10

Unit 13A page 38

1. 71, 89, 83, 64, 96, 80
2. 55, 43, 18, 36, 26, 62
3. 9, 36, 54, 72, 90, 63
4. 30, 9, 6, 18, 15, 27
5. 1, 9, 10, 11, 2, 8
6. 3, 8, 5, 9, 10, 6
7. 503
8. 12 106
9. 204
10. 5141
11. 1825
12. \$120
13. 31
14. 170
15. 7060
16. Th. H T U
17. 90c
18. \$20
19. 0.1 0.2 0.3 (arrow at 0.15)
20. 0.03 or 0.030
21. 207 000
22. 218 + 276 = 494

Unit 13B page 39

1. \$679 327
2. 170 674
3. 11 184
4. 184
5. $\frac{9}{10}$
6. 4
7. 24
8. 3:05 pm
9. 55 minutes
10. kg
11. 78
12. 1250 mL or 1.25 L
13. 8
14. 12
15. Henry
16. (1, 1) (4, 1) (2, 3) (1, 3)

17. right angle
18. 50% or $\frac{1}{2}$
19. 3
20. 11
21. parent/teacher to check
22. 30

Unit 14A page 40

1. 54, 82, 90, 36, 103, 77
2. 18, 8, 43, 24, 60, 49
3. 90, 18, 36, 81, 72, 54
4. 3, 27, 21, 33, 9, 15
5. 10, 4, 9, 5, 12, 3
6. 1, 11, 4, 7, 2, 10
7. 342
8. 9565
9. 323
10. 2395
11. 936
12. $57
13. 31
14. 230
15. 6985
16. Th. H T U
17. $2.45
18. $30
19. 0.81 (0.7, 0.8, 0.9)
20. 0.13 or 0.130
21. 9000
22. 798 – 529 = 269

Unit 14B page 41

1. $450 040
2. 733 552
3. 14 043
4. 167
5. $\frac{3}{5}$
6. 2
7. 25
8. 4:56 am
9. 45 minutes
10. g
11. 26
12. 1200 mL or 1.2 L
13. 3
14. 8
15. Joe
16. (1, 1) (3, 1) (4, 2) (3, 4) (1, 4)
17. straight angle
18. 0
19. 4 and 3
20. 18
21. parent/teacher to check
22. 15

Unit 15A page 42

1. 94, 64, 73, 45, 56, 77
2. 31, 46, 65, 32, 54, 79
3. 99, 18, 108, 63, 45, 27
4. 30, 24, 3, 12, 27, 18
5. 6, 4, 5, 9, 2, 10
6. 6, 10, 3, 2, 8, 9
7. 291
8. 7340
9. 312
10. 6656
11. 2856
12. $136
13. 62
14. 800
15. 28 729
16. Th. H T U
17. $1.20
18. $55
19. 0.55 (0.4, 0.5, 0.6)
20. 0.9 or 0.900
21. 50 000
22. 344 + 489 = 833

Unit 15B page 43

1. $90 894
2. 273 583
3. 58 328
4. 161
5. $\frac{2}{4} = \frac{1}{2}$
6. 1
7. 32
8. 9:20 pm
9. 1 hour 20 minutes
10. kg
11. 30
12. 1100 mL or 1.1 L
13. 4
14. 12
15. Phillip
16. (1, 2) (2, 0) (2, 4) (3, 2)
17. full revolution
18. $\frac{1}{4}$ or 25%
19. 7
20. 12
21. parent/teacher to check
22. 20

Revision 2A page 44

1. 75, 72, 54, 61, 81, 94
2. 29, 43, 65, 72, 1, 10
3. 36, 63, 81, 72, 90, 27
4. 7, 42, 35, 77, 84, 14
5. 12, 1, 4, 2, 5, 10
6. 12, 9, 8, 4, 6, 11
7. 733
8. 8545
9. 1435
10. A
11. $80
12. 3852
13. B
14. 900
15. B
16. 6302
17. $1.85
18. $42
19. B
20. C
21. 32 100
22. 3

Revision 2B page 45

1. $57 428
2. 210 000
3. 2100
4. false
5. $\frac{4}{10} = \frac{2}{5}$
6. $\frac{6}{18} = \frac{1}{3}$
7. A
8. 
9. 1:30 pm
10. B
11. 28
12. 725 mL
13. B
14. 8
15. apple
16. (1, 1) (4, 1) (4, 3) (1, 3)
17. A
18. $\frac{3}{8}$
19. 0
20. 18
21. parent/teacher to check
22. 150

NAPLAN-style Test 2 page 46

1. 4210
2. B
3. C
4. B
5. C
6. 4.25
7. A
8. B
9. C
10. $\frac{7}{10}$
11. B
12. C
13. B
14. C
15. B
16. (4, 3)
17. C
18. 10
19. D
20. C

Unit 16A page 50

1. 56, 49, 95, 82, 73, 60
2. 14, 44, 67, 31, 52, 70
3. 4, 24, 28, 32, 16, 36
4. 20, 50, 100, 30, 110, 120
5. 10, 9, 3, 7, 2, 1
6. 8, 1, 2, 3, 4, 6
7. 11 273
8. 513 m
9. $1100
10. 1123
11. 100, 180
12. 12 844
13. 400
14. 3 r 1 or $3\frac{1}{6}$
15. 42 119, 30 640
16. 43 256, 43 283, 43 295, 43 310
17. 2
18. $2.40
19. 0.08, 0.85, 0.90, 0.98
20. 14.35
21. 1000
22. 23

Unit 16B page 51

1. 90 000
2. 31 775
3. 2908
4. 161
5. A
6. $\frac{2}{5}$
7. 48
8. 6 hours
9. 7 : 15
10. 6000
11. 40 cm
12. t
13.
14. rectangle, triangle, rectangle
15. Todd River
16. Olive Pink Botanic Park
17. obtuse
18. apple
19. parent/teacher to check
20. 6
21. 20
22. parent/teacher to check

Unit 17A page 52

1. 83, 38, 100, 66, 71, 92
2. 18, 25, 64, 5, 40, 39
3. 8, 32, 20, 40, 0, 44
4. 10, 90, 60, 40, 70, 30
5. 2, 5, 4, 6, 10, 9
6. 10, 2, 9, 3, 4, 11
7. 8177
8. 506 kg
9. $1650

10. 9118
11. 240, 280
12. 3102
13. 206
14. 6 r 2 or $6\frac{2}{3}$
15. 21 953, 33 476
16. 156 211, 156 860, 157 004, 157 398
17. 10
18. $1.10
19. 0.30, 0.38, 0.83, 0.88
20. 1.29
21. 2900
22. 31

Unit 17B page 53

1. 110 000
2. 462 509
3. 5464
4. 112
5. B
6. $\frac{4}{8} = \frac{1}{2}$
7. 132
8.  
9. 1 hour 5 minutes
10. 4000
11. 70 cm
12. g
13. 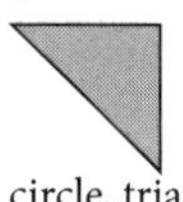
14. circle, triangle, triangle
15. Alice Springs Desert Park
16. Telegraph Station Historic Reserve
17. acute
18. banana
19. parent/teacher to check
20. 16
21. 75
22. parent/teacher to check

Unit 18A page 54

1. 96, 71, 100, 99, 62, 84
2. 84, 39, 50, 77, 11, 25
3. 0, 12, 24, 36, 44, 20
4. 20, 70, 80, 10, 100, 40
5. 12, 6, 4, 8, 5, 11
6. 2, 10, 5, 4, 6, 11
7. 6731
8. 1950 mm
9. $2647
10. 1089
11. 360, 480
12. 6510
13. 903
14. 4 r 2 or $4\frac{2}{7}$
15. 78 399, 50 246
16. 262 109, 265 327, 266 538, 267 854
17. 4
18. $5.70
19. 0.28, 0.36, 0.4, 0.5
20. 13.49
21. 4100
22. 8 or 42

Unit 18B page 55

1. 160 000
2. 68 554
3. 1192
4. 66
5. A
6. $\frac{5}{10} = \frac{1}{2}$
7. 88
8.
9. 3 hours 30 minutes
10. 1000
11. 90 cm
12. kg
13. 
14. rectangle, square, rectangle
15. Anzac Hill
16. Alice Springs Desert Park
17. obtuse
18. banana
19. parent/teacher to check
20. 69
21. 150
22. parent/teacher to check

Unit 19A page 56

1. 100, 42, 75, 49, 58, 92
2. 41, 65, 26, 52, 34, 80
3. 12, 4, 32, 36, 8, 16
4. 50, 70, 90, 110, 120, 60
5. 7, 3, 9, 5, 8, 10
6. 12, 1, 9, 2, 6, 4
7. 10 794
8. 1131 g
9. $4107
10. 4157
11. 210, 540
12. 3410
13. 606
14. 5 r 5 or $5\frac{5}{9}$
15. 99 466, 28 430
16. 4 220 358, 4 225 717, 4 226 843, 4 230 109
17. 2
18. $3.85
19. 0.46, 0.59, 0.64, 0.72
20. 11.04
21. 7300
22. 23

Unit 19B page 57

1. 100 000
2. 569 732
3. 4158
4. 162
5. B
6. $\frac{2}{6} = \frac{1}{3}$
7. 202
8.
9. 1 hour 50 minutes
10. 5000
11. 110 cm
12. g
13. 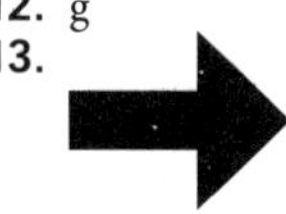
14. square, triangle, triangle
15. Alice Springs Desert Park
16. a road
17. reflex
18. orange
19. parent/teacher to check
20. 100
21. 1400
22. parent/teacher to check

Unit 20A page 58

1. 77, 61, 93, 70, 68, 59
2. 64, 59, 46, 20, 37, 51
3. 8, 24, 28, 48, 16, 44
4. 10, 50, 80, 90, 110, 30
5. 10, 4, 12, 5, 2, 11
6. 6, 9, 3, 1, 11, 4
7. 6393
8. 1336 L
9. $7071
10. 1164
11. 180, 720
12. 4688
13. 603
14. 6 r 4 or $6\frac{4}{6}$ or $6\frac{2}{3}$
15. 246 381, 380 571
16. 2 082 463, 2 083 498, 2 085 762, 2 085 775
17. 4
18. 50c
19. 0.11, 0.19, 0.2, 0.5
20. 13.35
21. 3000
22. 68 (or 102)

Unit 20B page 59

1. 210 000
2. 170 239
3. 4968
4. 171
5. B
6. $\frac{2}{10} = \frac{1}{5}$
7. 420
8. 
9. 2 hours 30 minutes
10. 9000
11. 104 cm
12. t
13. 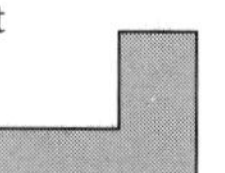
14. circle, rectangle, rectangle
15. Telegraph Station Historic Reserve
16. Olive Pink Botanic Park
17. reflex
18. $\frac{3}{10}$ or 0.3 or 30%
19. parent/teacher to check
20. 60
21. sport
22. parent/teacher to check

Unit 21A page 60

1. 146, 185, 192, 138, 140, 179
2. 50, 185, 262, 196, 1, 250
3. 20, 60, 30, 80, 70, 50
4. 3, 27, 30, 33, 12, 36
5. 11, 4, 9, 6, 1, 10
6. 11, 4, 2, 9, 3, 10
7. 36 455
8. 30 000
9. 1214
10. $727
11. 108
12. 3, 6, 9, 12, 15
13. 1, 2, 4, 5, 10, 20
14. 10
15. 8 433 210
16. two hundred and four thousand, six hundred and thirty-eight
17. $8.50
18. $2.45
19. 0.84
20. 43
21. 960
22. 40, 56, 56, 61

Unit 21B page 61

1. 1 190 397
2. 89 464
3. 1125
4. 1613
5. $\frac{3}{4}$
6. $\frac{3}{3} = 1$
7. 81
8. A 7:45, B 8:15, C 8:30
9. 1:08 am
10. 2 kg 750 g
11. 15
12. 5, 1, 1, 5
13. 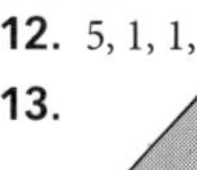
14. D

15. parent/teacher to check
16. parent/teacher to check
17. 90°
18. 0.5
19. R, G, W, Y
20.

Dice number	Number
1	5
2	4
3	4
4	5

21. parent/teacher to check
22. raspberry

Unit 22A page 62

1. 526, 344, 238, 299, 302, 639
2. 363, 146, 45, 187, 211, 7
3. 100, 60, 110, 20, 70, 90
4. 20, 60, 15, 40, 25, 5
5. 2, 9, 3, 5, 8, 1
6. 10, 4, 8, 5, 9, 3
7. 106 878
8. 53 000
9. 1454
10. 168 m
11. 224
12. 4, 8, 12, 16, 20
13. 1, 2, 4, 5, 8, 10, 20, 40
14. 5
15. 9 632 210
16. two million, one hundred and seventy thousand, eight hundred and fifty
17. $4.05
18. $6.75
19. 1.35
20. 79
21. 980
22. 36, 18, 43, 33

Unit 22B page 63

1. 1 153 576
2. 194 032
3. 2569
4. 807
5. $\frac{5}{6}$
6. $\frac{3}{5}$
7. 178
8. A 2:15, B 12:45, C 1:30
9. 9:00 pm
10. 1 kg 160 g
11. 9
12. 2, 2, 2, 8
13.
14. C
15. parent/teacher to check
16. parent/teacher to check
17. 40°
18. $\frac{3}{8}$
19. 1, 2, 3, 4
20.

Shape	Number
square	5
triangle	6
circle	11

21. parent/teacher to check
22. May

Unit 23A page 64

1. 571, 419, 785, 608, 325, 499
2. 285, 363, 2, 473, 119, 250
3. 70, 60, 30, 80, 20, 90
4. 28, 70, 77, 35, 84, 0
5. 4, 6, 2, 9, 8, 3
6. 10, 4, 1, 3, 7, 6
7. 89 585
8. 51 000
9. 1257
10. 142
11. 108
12. 6, 12, 18, 24, 30
13. 1, 2, 3, 4, 6, 8, 12, 24
14. 10
15. 9 763 211
16. two million, sixteen thousand, two hundred and nine
17. $7.50
18. $1.20
19. 5.09
20. 86
21. 1340
22. 31, 46, 84, 46

Unit 23B page 65

1. 897 838
2. 516 176
3. 2232
4. 1076
5. $\frac{7}{8}$
6. $\frac{5}{8}$
7. 92
8. A 11:45, B 12:15, C 12:00
9. 2:40 pm
10. 4 kg 700 g
11. 7
12. 3, 2, 3, 18
13. 
14. B
15. parent/teacher to check
16. parent/teacher to check
17. 70°
18. 0.25
19. R, Y, O
20.

Shape	Number
star	10
moon	6
sun	6

21. parent/teacher to check
22. Tuesday

Revision 3A page 66

1. 36, 44, 112, 55, 90, 83
2. 68, 57, 22, 10, 39, 48
3. 28, 36, 12, 48, 4, 40
4. 16, 40, 64, 32, 88, 48
5. 11, 4, 7, 8, 9, 2
6. 10, 3, 6, 5, 2, 12
7. B
8. C
9. 11 300
10. 415
11. <
12. 28 788
13. 1, 2, 3, 6, 9, 18
14. 3 r 3 or $3\frac{3}{6}$ or $3\frac{1}{2}$
15. D
16. one million, seven hundred and sixty-five thousand, two hundred
17. $12.20
18. 4
19. 11.42
20. 34.4 kg
21.

Number	Round to the nearest 10	Round to the nearest 100
3216	3220	3200
40 793	40 790	40 800
118 306	118 310	118 300

22. 79

Revision 3B page 67

1. 245 000
2. 274 169
3. false
4. 132
5. B
6. $\frac{3}{5}$
7. 152
8. C
9. 2 hours 45 minutes
10.

Mass	kg	g
4250 g	4	250
3060 g	3	60
1490 g	1	490

11. 15
12. 5, 2, 2, 20
13. 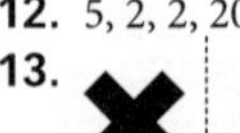 

14. 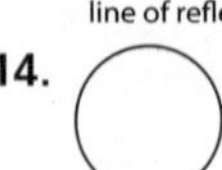
15. sun
16.

17. 160°
18. $\frac{3}{7}$
19. 0.5
20.

Weather	No. of days
sun	8
rain	4
fog	2
cloud	6

21. 30
22. parent/teacher to check

NAPLAN-style Test 3 page 68

1. D
2. A
3. D
4. C
5. B
6. C
7. B
8. 1.22
9. D
10. B
11. C
12. <
13. C
14. C
15. B
16. A
17. C
18. A
19. 40 + 50 + 30 = 120
20. B

Unit 24A page 72

1. 618, 866, 510, 725, 572, 648
2. 11, 160, 220, 287, 79, 108
3. 100, 50, 80, 70, 20, 110
4. 18, 54, 24, 72, 36, 0
5. 10, 6, 7, 4, 1, 11
6. 8, 11, 3, 5, 2, 7
7. 40 220
8. 35 000
9. 4682
10. 696
11. 150
12. 7, 14, 21, 28, 35
13. 1, 2, 3, 5, 6, 10, 15, 30
14. 12
15. 9 864 321
16. six million, forty-two thousand, one hundred and seventy-nine
17. $6.40
18. $3.65
19. 1.63
20. 64
21. 1040
22. 52, 14, 52, 63

Unit 24B page 73

1. 1 259 748
2. 287 649
3. 3234
4. 1613
5. $\frac{9}{10}$
6. $\frac{9}{10}$
7. 43
8. A 7:45, B 8:30, C 8:00
9. 1.50 am
10. 3 kg 950 g
11. 12
12. 2, 1, 5, 10
13.
14. E
15. parent/teacher to check
16. parent/teacher to check
17. 120°
18. 0
19. 1, 2, 3, 4, 6, 8
20.

Colour	Number
red	9
green	7
blue	5
white	7
yellow	6

21. parent/teacher to check
22. 5

Unit 25A page 74

1. 523, 209, 338, 449, 387, 277
2. 101, 527, 335, 755, 480, 363
3. 60, 30, 90, 10, 80, 70
4. 32, 80, 16, 88, 96, 40
5. 12, 1, 6, 5, 2, 9
6. 8, 10, 2, 6, 3, 9
7. 50 012
8. 4300
9. 2297
10. 1165
11. 112
12. 3, 6, 9, 12, 15
13. 1, 2, 3, 6, 9, 18
14. 9
15. 9 875 431
16. six million, one hundred and eighty thousand, three hundred and twenty-one
17. $11.75
18. $6.10
19. 4.22
20. 86
21. 1200
22. 85, 9, 67, 37

Unit 25B page 75

1. 701 092
2. 785 568
3. 2910
4. 2312
5. $\frac{2}{3}$
6. $\frac{5}{6}$
7. 193
8. A 10:30, B 10:15, C 10:00
9. 4:15 am
10. 5 kg 900 g
11. 14
12. 3, 4, 3, 36
13.
14. A
15. parent/teacher to check
16. parent/teacher to check
17. 100°
18. $\frac{1}{8}$
19. 9, 4, 6
20.

Vehicle	Number
car	8
plane	4
truck	6

21. parent/teacher to check
22. 300

Unit 26A page 76

1. 45, 53, 64, 34, 115, 102
2. 8, 27, 66, 53, 72, 31
3. 400, 700, 800, 900, 600, 300
4. 8, 88, 40, 80, 96, 16
5. 8, 1, 7, 2, 4, 9
6. 7, 5, 9, 3, 8, 10
7. 3256 + 1526 = 4782
8. 6116
9. 36 191
10. 1245
11. 8, 16, 24, 32, 40
12. 36
13. 92.5
14. 1, 2, 4, 8, 16
15. 143
16. <
17. $18.53, $18.55
18. $24.40
19. 1.35
20. 3.3
21. $14
22. 71

Unit 26B page 77

1. 1 015 788
2. 130 000
3. 182
4. 1196
5. $\frac{1}{8}$ 6. $\frac{4}{5}$
7. 102
8. 11:55 am
9.
10. 120 cm
11. 8
12. 290 g
13. 2
14. parallelogram
15. parent/teacher to check
16. 2
17. 30°
18. parent/teacher to check
19. $\frac{3}{8}$
20.

Colour	Number
gold	10
silver	7
bronze	15

21. $\frac{1}{2}$
22. parent/teacher to check

Unit 27A page 78

1. 57, 62, 109, 91, 43, 39
2. 9, 54, 25, 62, 30, 74
3. 110, 90, 20, 30, 40, 70
4. 6, 36, 60, 48, 72, 30
5. 5, 6, 8, 2, 3, 10
6. 6, 0, 3, 9, 1, 4
7. 4638 + 3121 = 7759
8. 5826
9. 47 737
10. 7033
11. 6, 12, 18, 24, 30
12. 64
13. 64.5
14. 1, 2, 5, 10, 25, 50
15. 217
16. <
17. $5.62, $5.60
18. $36.30
19. 7.28
20. 4.1
21. $14
22. 36

Unit 27B page 79

1. 801 791
2. 310 000
3. 216
4. 1004
5. $\frac{1}{10}$
6. $\frac{2}{8} = \frac{1}{4}$
7. 12
8. 8:10 am
9. 
10. 1 km
11. 9
12. 500 g
13. 5
14. trapezium
15. parent/teacher to check
16. 1
17. 130°
18. parent/teacher to check
19. $\frac{2}{8} = \frac{1}{4}$
20.

Day	Number
Monday	10
Tuesday	30
Wednesday	20
Thursday	40

21. 10 °C
22. parent/teacher to check

Unit 28A page 80

1. 40, 64, 86, 97, 68, 53
2. 53, 62, 1, 25, 40, 79
3. 400, 1000, 900, 1100, 300, 700
4. 9, 54, 108, 45, 72, 18
5. 7, 6, 9, 5, 1, 3
6. 3, 6, 9, 4, 10, 5
7. 4063 + 3845 = 7908
8. 16 080
9. 60 532
10. 5392
11. 7, 14, 21, 28, 35
12. 100
13. 32.9
14. 1, 2, 3, 4, 6, 8, 12, 16, 24, 48
15. 163
16. >
17. $1.61, $1.60
18. $42.40
19. 17.92
20. 3.1
21. $18
22. 125

Unit 28B page 81

1. 832 321
2. 27 000
3. 504
4. 483
5. $\frac{1}{5}$
6. $\frac{9}{10}$
7. 95
8. 9:05 am
9. 
10. 1 km
11. 48
12. 1100 g
13. 2
14. hexagon
15. parent/teacher to check
16. 4
17. 180°
18. parent/teacher to check
19. $\frac{3}{8}$

20.

Size	Number
small	150
medium	200
large	100

21. 10
22. parent/teacher to check

Unit 29A page 82

1. 71, 108, 41, 98, 84, 58
2. 25, 50, 24, 59, 33, 11
3. 700, 400, 800, 900, 100, 300
4. 15, 30, 33, 0, 18, 6
5. 10, 3, 1, 2, 5, 9
6. 5, 11, 3, 7, 1, 12
7. 1676 + 4383 = 6059
8. 14 449
9. 34 157
10. 3947
11. 4, 8, 12, 16, 20
12. 49
13. 78.3
14. 1, 3, 5, 9, 15, 45
15. 321
16. <
17. $2.51, $2.50
18. $35.95
19. 7.37
20. 8.0
21. $10
22. 492

Unit 29B page 83

1. 1 023 234
2. 45 000
3. 4158
4. 667
5. $\frac{1}{8}$
6. $\frac{6}{8} = \frac{3}{4}$
7. 135
8. 6:35 pm
9. 
10. 200 cm
11. 100
12. 1400 g
13. 1
14. rectangle
15. parent/teacher to check
16. 3
17. 60°
18. parent/teacher to check
19. stripey
20.

Pet	Number
dog	10
cat	20
bird	30
rabbit	25

21. $\frac{1}{4}$
22. parent/teacher to check

Unit 30A page 84

1. 41, 30, 65, 79, 19, 58
2. 14, 30, 45, 37, 66, 13
3. 700, 500, 900, 400, 1100, 300
4. 14, 70, 84, 42, 7, 56
5. 7, 5, 3, 2, 9, 4
6. 1, 4, 7, 9, 6, 2
7. 5363 + 2468 = 7831
8. 4660
9. 23 537
10. 1627
11. 12, 24, 36, 48, 60
12. 25
13. 91.2
14. 1, 2, 5, 7, 10, 14, 35, 70
15. 1216
16. <
17. $10.12, $10.10
18. $13.95
19. 6.97
20. 10.0
21. $3
22. 532

Unit 30B page 85

1. 703 238
2. 25 000
3. 2288
4. 1032
5. $\frac{1}{8}$
6. $\frac{3}{6} = \frac{1}{2}$
7. 53
8. 7:38 pm
9. 
10. 1 m
11. 36
12. 1000 g
13. 1
14. octagon
15. parent/teacher to check
16. 2
17. 85°
18. parent/teacher to check
19. spotty fish
20.

Insects	Number
fly	40
bee	20
cricket	10
ant	60

21. 50
22. parent/teacher to check

Revision 4A page 86

1. 42, 50, 63, 46, 69, 83
2. 11, 49, 30, 32, 53, 71
3. 60, 70, 40, 110, 10, 30
4. 12, 48, 30, 60, 72, 54
5. 4, 9, 11, 10, 3, 7
6. 12, 10, 5, 6, 2, 1
7. 4703 – 479 = 4224
8. 98 000
9. 7328
10. 2003
11. 90
12. C
13. 1, 2, 3, 5, 6, 10, 15, 30
14. 32.7, 3.27
15. 613
16. >
17. $7.05
18. C
19. A
20. 0.09, 0.60, 0.67, 0.76
21. $16.80
22. 39, 46, 46, 32, 34

Revision 4B page 87

1. 840 971
2. 650 000
3. 2492
4. 1538
5. D
6. $\frac{2}{8} = \frac{1}{4}$
7. 35
8. C
9. 
10. C
11. 32
12. C
13. 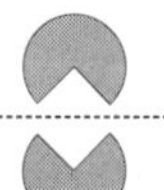

14. B
15. parent/teacher to check
16. square
17. 100°
18. spots, wide stripes, shading, plain, thin stripes
19. $\frac{5}{10} = \frac{1}{2}$
20.

Equipment	Number
swing	15
slide	10
seesaw	20
monkey bars	5

21. $\frac{1}{4}$
22. 4

NAPLAN-style Test 4 page 88

1. D
2. 86 936
3. D
4. D
5. B
6. B
7. B
8. C
9. C
10. A
11. C
12. C
13. C
14. C
15. D
16. A
17. D
18. 32
19. B
20. C

6 Ben bought a roll for lunch.
He had \$3.50 and borrowed another 75c to pay for the roll.
How much did the roll cost?

\$ ☐

7 What value is the arrow pointing to?

A 7.45
B 7.5
C 7.05
D 7.15

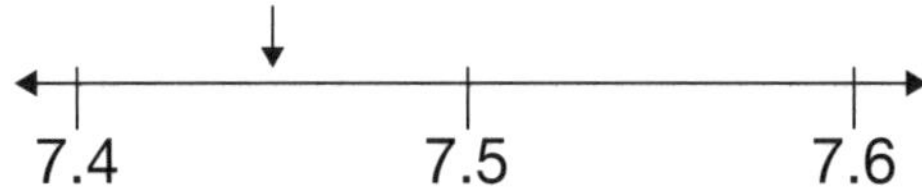

8 What is the missing number?

$$13 \times 6 = \square \times 26$$

A 2
B 3
C 9
D 12

9 What is the difference between the prices of the houses?

A \$394 395
B \$384 304
C \$384 395
D \$385 395

10

$$\frac{3}{10} + \frac{4}{10} = \square$$

11 Tracey wants to leave home 1 hour before the movie.
The movie starts at 3:45 pm. What time does Tracey want to leave?

A 4:45 pm
B 2:45 pm
C 2:15 pm
D 3:00 pm

12 The garden bed is a rectangle that is 5 m long and 2 m wide.
What is the perimeter of the garden bed?

A 7 m
B 12 m
C 14 m
D 10 m

13 Henry needs to take 1000 mL of water for his hike. He has one container which holds 250 mL of water.
How much more water does he need to take?

A 250 mL
B 750 mL
C 850 mL
D 1250 mL

14 A dice is shaped like a cube. How many edges does a dice have?

A 4
B 8
C 12
D 16

15 The door is opened at an angle of 120°. What type of angle is this?

A acute
B obtuse
C reflex
D straight

16 Sara is drawing a square on the grid.
What are the coordinates of the fourth corner of the square? (______, _______)

---

17 Here are 7 cards. Which card is the most likely to be selected?

 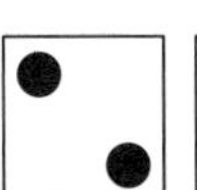 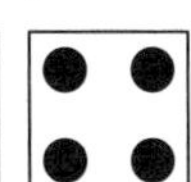 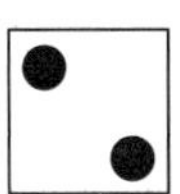 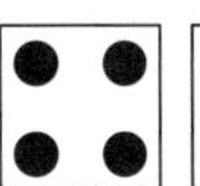 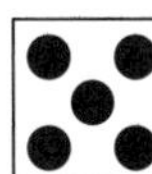

A

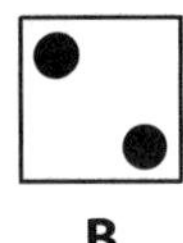
B

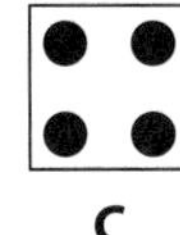
C

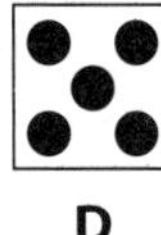
D

---

18 This table relates to the graph.
What is the missing value of the D batteries?

Battery type	Number
AA	15
C	23
D	
9V	20

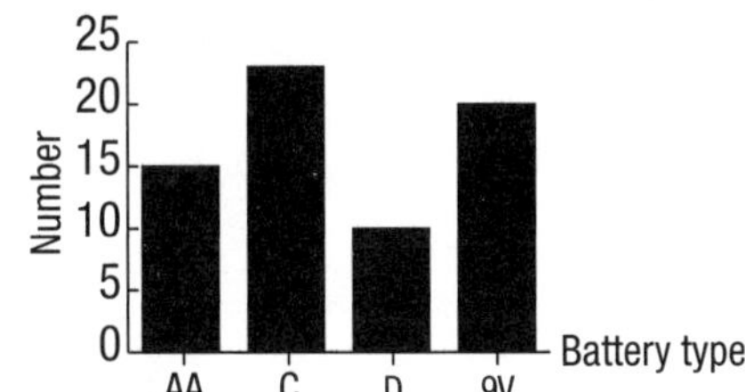

---

19 What is the missing number?

$$44 + 33 - 12 = (5 \times \square) + 5$$

A 13
B 17
C 10
D 12

---

20 The farmer puts 2 pineapples and 5 mangoes in every fruit box.
How many pineapples does the farmer need for 6 boxes?

A 3
B 6
C 12
D 42

UNIT 16A

1

+	46	39	85	72	63	50
10						

2

–	34	64	87	51	72	90
20						

3

×	1	6	7	8	4	9
4						

4

×	2	5	10	3	11	12
10						

5

÷	70	63	21	49	14	7
7						

6

÷	80	10	20	30	40	60
10						

7

$$\begin{array}{r} 8356 \\ +\ 2917 \\ \hline \end{array}$$

8 What is the total length of 156 m, 236 m and 121 m?

9 What is the difference between \$900 and \$2000?

10 There are 1421 toys in a warehouse. 298 are delivered to a shop. How many toys are left?

11

$5 \times 20 =$ ☐

$6 \times 30 =$ ☐

12

$$\begin{array}{r} 3211 \\ \times\ \ 4 \\ \hline \end{array}$$

13 Find the missing digits.

$$\begin{array}{r} 100 \\ 4\overline{)\square\square\square} \end{array}$$

14

$19 \div 6 =$ ☐

15 Circle the smaller number in each pair.

42 119   43 103   30 670   30 640

16 Order the numbers from smallest to largest.

43 295   43 310   43 256   43 283

17 How many drinks can Stan buy with \$5.00, if one drink costs \$2.50?

18 The shopping cost \$7.60. What is Dad's change from \$10.00?

19 Order the decimals from smallest to largest.

0.98   0.08   0.90   0.85

20 Add 7.95 and 6.4.

21 Round 1046 to the nearest hundred.

22 27 and another number are added to make 50. What is the other number?

1 $20\,000 + 40\,000 + 30\,000 =$ ☐

2 

$81\,738 - 49\,963 =$ ☐

3 Find the product of 727 and 4.

☐

4 Share 966 oranges equally between 6 bags. How many oranges will be in each bag?

☐

5 Circle the larger shaded fraction.

A 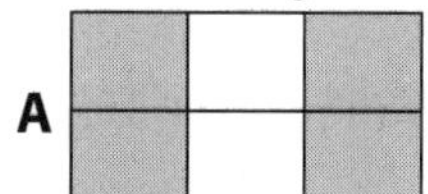 B 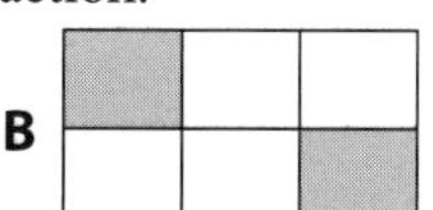

6 Use the diagram to find:

$\frac{3}{5} - \frac{1}{5} =$ ☐

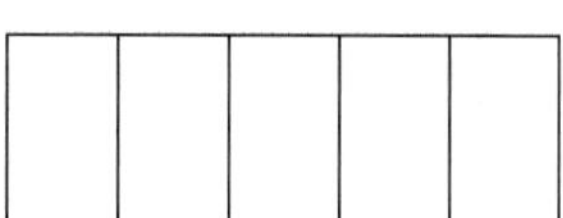

7 Sophie has 11 cards. Cindy has 6 cards more than Sophie and Sean has 9 more cards than Sophie. What is the total number of cards?

☐

8 Judy starts school at 9:00 am and finishes at 3:00 pm. How long is Judy at school?

☐

9 Write a quarter past seven on the clock.

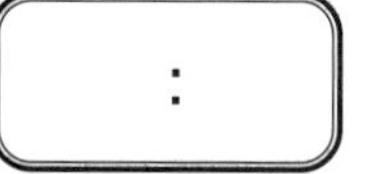

10 How many metres is 6 km?

☐ m

11 Complete the table.

Length	Width	Perimeter
8 cm	12 cm	

12 Tick to select the most appropriate unit to measure the mass.

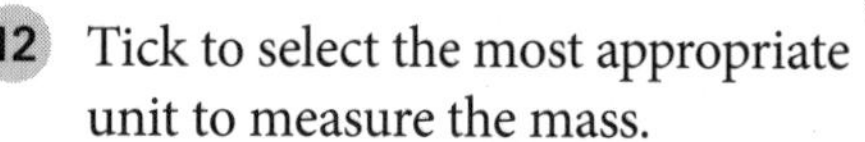

Object	g	kg	t
mass of a train			

13 Draw the triangle rotated a quarter turn anticlockwise.

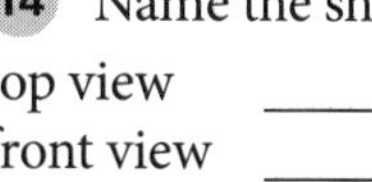

14 Name the shape of each of the views.

top view __________

front view __________

side view __________

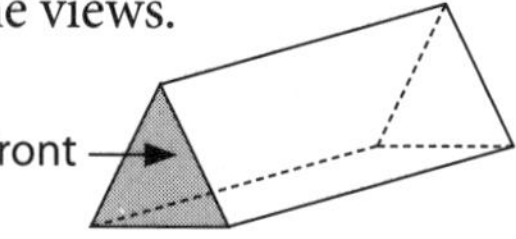

15 What river is to the east of Alice Springs?

☐

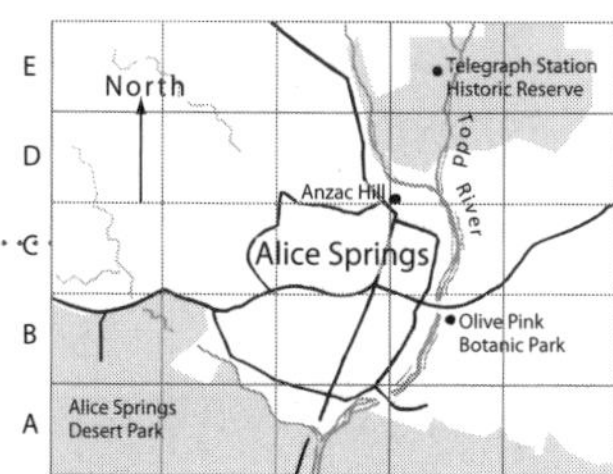

16 What is at 4B?

☐

17 Circle the angle type for 105°.

straight obtuse acute reflex right

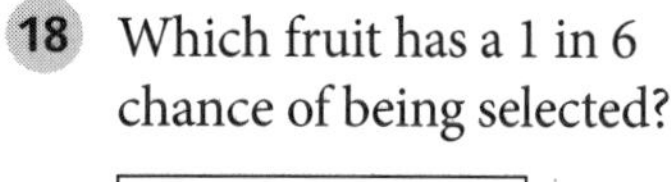

18 Which fruit has a 1 in 6 chance of being selected?

☐

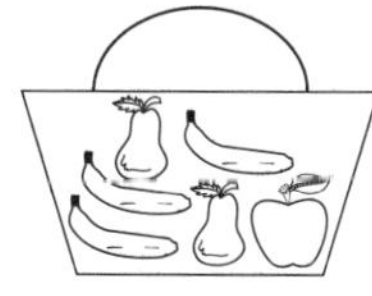

19 Mark the word **unlikely** on the scale.

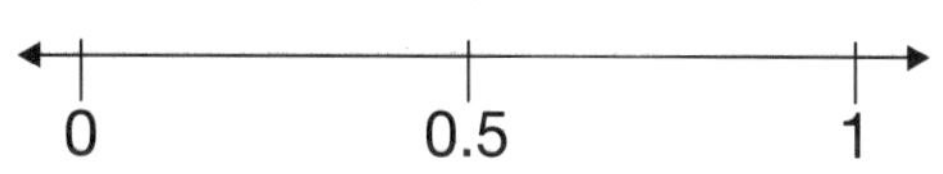

20 How many people followed the silver team?

☐

Team colour	Number
black	8
green	7
red	21
silver	6

21 How many people prefer thongs?

☐

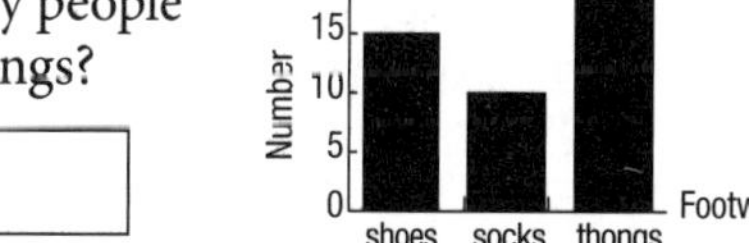

22 Create a dot plot for the following data.

No. of pets	No. of people
0	3
1	6
2	7
3	4

1

+	63	18	80	46	51	72
20						

2

−	48	55	94	35	70	69
30						

3

×	2	8	5	10	0	11
4						

4

×	1	9	6	4	7	3
10						

5

÷	14	35	28	42	70	63
7						

6

÷	100	20	90	30	40	110
10						

7

$$\begin{array}{r} 6394 \\ +\ 1783 \\ \hline \end{array}$$

8 What is the total mass of 129 kg, 242 kg and 135 kg?

9 What is the difference between $2700 and $1050?

10 There are 9335 apples in the warehouse. 217 are delivered to a shop. How many apples are left?

11

$8 \times 30 = \square$

$7 \times 40 = \square$

12

$$\begin{array}{r} 1034 \\ \times\ \ 3 \\ \hline \end{array}$$

13 Find the missing digits.

$$\begin{array}{r} 103 \\ 2\overline{)\square\square\square} \end{array}$$

14

$20 \div 3 = \square$

15 Circle the smaller number in each pair.

21 953 22 184   33 476 39 100

16 Order the numbers from smallest to largest.

156 211 157 398 156 860 157 004

17 How many pears can Pam buy with $5.00, if one pear costs 50c?

18 Mum bought some vegetables for $8.90. What was her change from $10.00?

19 Order the decimals from smallest to largest.

0.83 0.38 0.88 0.30

20 Add 0.38 and 0.91.

21 Round 2930 to the nearest hundred.

22 39 and another number are added to make 70. What is the other number?

1 $40\,000 + 10\,000 + 60\,000 = \square$

2 $738\,946 - 276\,437 = \square$

3 Find the product of 683 and 8.

4 Share 896 stamps equally between 8 containers. How many stamps will be in each container?

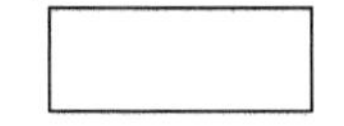

5 Circle the larger shaded fraction.

A 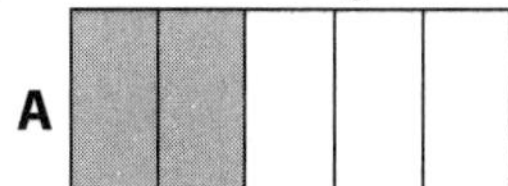

B 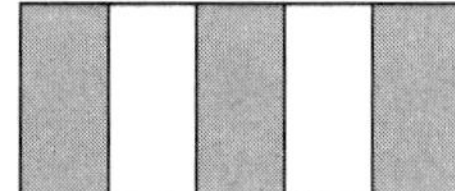

6 Use the diagram to find:

$\frac{7}{8} - \frac{3}{8} = \square$

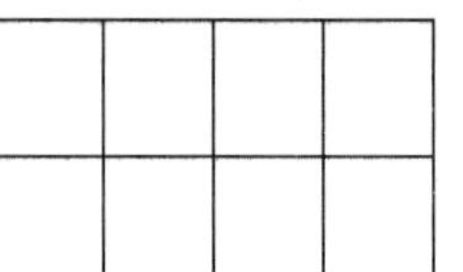

7 There are 9 paddocks with 10 sheep in each and 7 paddocks with 6 sheep in each. How many sheep are there altogether?

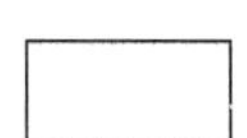

8 Draw 2:05 on the clock.

9 Jon starts lunch at 11:45 am and finishes at 12:50 pm. How long is Jon's lunch break?

10 How many metres is 4 km?

$\square$ m

11 Complete the table.

Length	Width	Perimeter
20 cm	15 cm	

12 Tick to select the most appropriate unit to measure the mass.

Object	g	kg	t
mass of an ant			

13 Draw the triangle rotated half a turn clockwise.

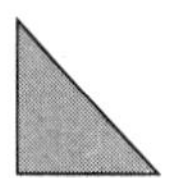

14 Name the shape of each of the views.

top view ____________

front view ____________

side view ____________

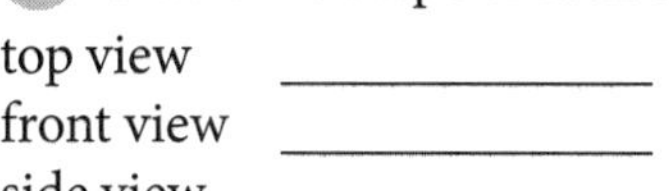

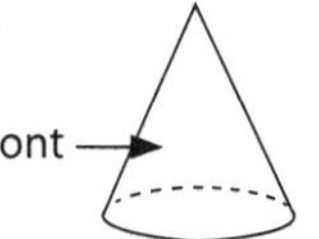

15 What is the name of the park to the south-west of Alice Springs?

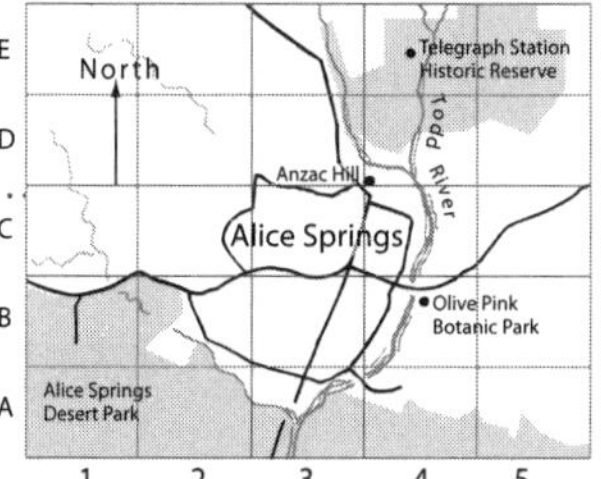

16 What is at 4E?

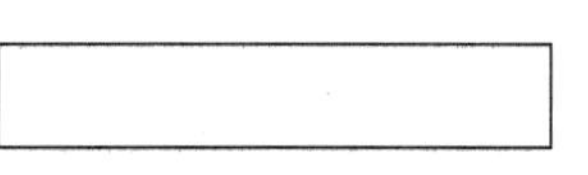

17 Circle the angle type for 80°.

straight   obtuse   acute   reflex   right

18 Which fruit is the most likely to be selected?

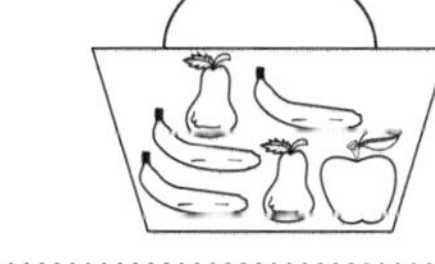

19 Mark the words **equal chance** on the scale.

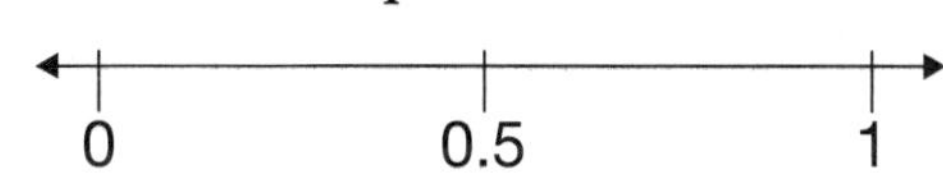

20 How many small tops have been ordered?

Size	Number
small	16
medium	9
large	12

21 What is the number of plums?

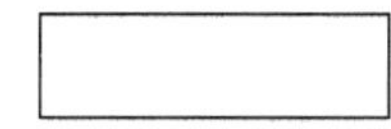

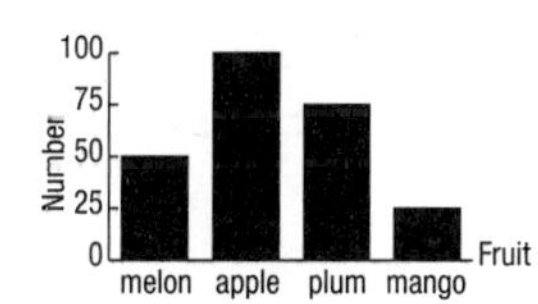

22 Create a dot plot for the following data.

No. of people in family	No. of people surveyed
3	9
4	8
5	6
6	3

1

+	66	41	70	69	32	54
30						

2

−	94	49	60	87	21	35
10						

3

×	0	3	6	9	11	5
4						

4

×	2	7	8	1	10	4
10						

5

÷	84	42	28	56	35	77
7						

6

÷	20	100	50	40	60	110
10						

7

$$\begin{array}{r} 4783 \\ +\ 1948 \\ \hline \end{array}$$

8 What is the total length of 683 mm, 926 mm and 341 mm?

9 What is the difference between $1725 and $4372?

10 There are 1476 cabbages in the field. Then 387 are picked. How many cabbages are left?

11

9 × 40 = ☐

8 × 60 = ☐

12

$$\begin{array}{r} 1302 \\ \times\ \ \ 5 \\ \hline \end{array}$$

13 Find the missing digits.

$$\begin{array}{r} 301 \\ 3\,\overline{)\,\square\square\square} \end{array}$$

14

30 ÷ 7 = ☐

15 Circle the smaller number in each pair.

79 615 78 399 50 246 50 392

16 Order the numbers from smallest to largest.

266 538 262 109 267 854 265 327

17 How many cupcakes can Lisa buy with $5.00 if one cupcake costs $1.20?

18 Linda buys fruit costing $4.30. What is her change from $10.00?

19 Order the decimals from smallest to largest.

0.4 0.36 0.28 0.5

20 Add 6.1 and 7.39.

21 Round 4096 to the nearest hundred.

22 The difference between a number and 25 is 17. What is the number?

1 $30\,000 + 70\,000 + 60\,000 = \square$

2 $368\,421 - 299\,867 = \square$

3 Find the product of 298 and 4.

4 Share 462 nails equally between 7 containers. How many nails will be in each container?

5 Circle the larger shaded fraction.

A 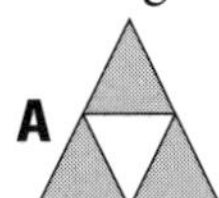

B 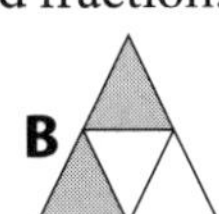

6 Use the diagram to find:

$\frac{8}{10} - \frac{3}{10} = \square$

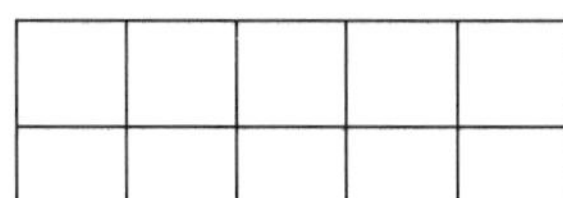

7 There are 8 biscuits in 8 packets and 6 biscuits in 4 packets. How many biscuits are there altogether?

8 Draw 9:35 on the clock.

9 The swimming carnival starts at 11:00 am and finishes at 2:30 pm. How long does the swimming carnival go for?

10 How many metres is 1 km?

□ m

11 Complete the table.

Length	Width	Perimeter
30 cm	15 cm	

12 Tick to select the most appropriate unit to measure the mass.

Object	g	kg	t
mass of a bowling ball			

13 Draw the shape rotated a quarter turn clockwise.

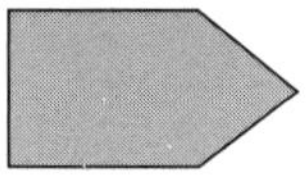

14 Name the shape of each of the views.

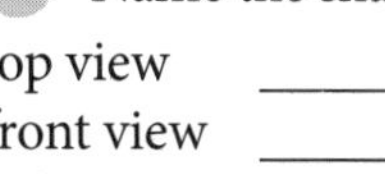

top view ____________

front view ____________

side view ____________

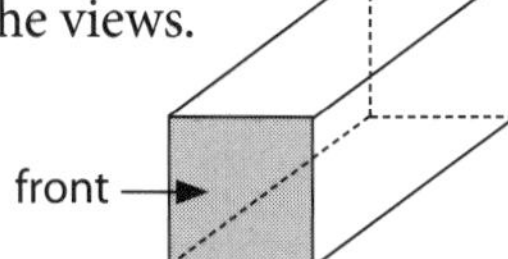

15 What is the name of the hill to the north of Alice Springs?

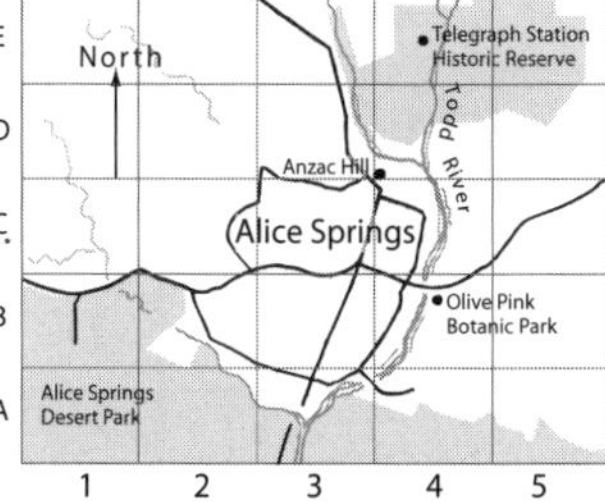

16 What is at 1A?

17 Circle the angle type for 115°.

straight obtuse acute reflex right

18 Which fruit has 50% chance of being selected?

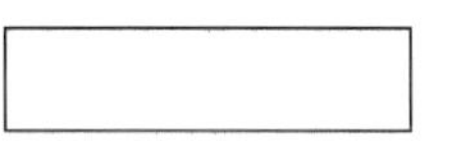

19 Mark the word **possible** on the scale.

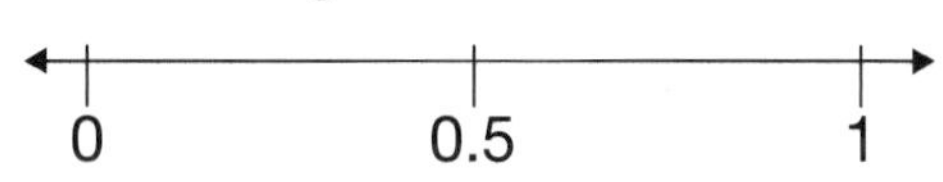

20 What is the total number of pieces of cutlery?

Cutlery	Number
knife	20
fork	23
spoon	26

21 What is the number of silver medals won?

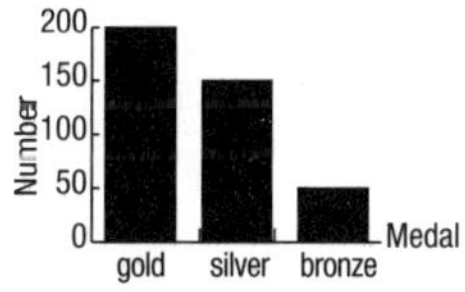

Number: 0, 50, 100, 150, 200; Medal: gold, silver, bronze

22 Create a dot plot for the following data.

No. of TVs	No. of people surveyed
1	6
2	10
3	15
4	4

1

+	60	2	35	9	18	52
40						

2

–	61	85	46	72	54	100
20						

3

×	3	1	8	9	2	4
4						

4

×	5	7	9	11	12	6
10						

5

÷	49	21	63	35	56	70
7						

6

÷	120	10	90	20	60	40
10						

7

$$\begin{array}{r} 7938 \\ +\ 2856 \\ \hline \end{array}$$

8 What is the total mass of 142 g, 689 g and 300 g?

9 What is the difference between $4063 and $8170?

10 There are 5376 jelly beans at the factory 1219 are sold to a shop. How many jelly beans are left?

11

$7 \times 30 =$ ☐

$9 \times 60 =$ ☐

12

$$\begin{array}{r} 1705 \\ \times\ \ \ 2 \\ \hline \end{array}$$

13 Find the missing digits.

$$6\,\overline{)\square\square\square}\quad = 101$$

14

$50 \div 9 =$ ☐

15 Circle the smaller number in each pair.

110 325   99 466   28 430   29 773

16 Order the numbers from smallest to largest.

4 220 358   4 230 109   4 225 717   4 226 843

17 How many ice creams can Paul buy with $5.00 if one ice cream costs $1.80?

18 Linda buys bread costing $6.15. What is her change from $10.00?

19 Order the decimals from smallest to largest.

0.46   0.64   0.59   0.72

20 Add 7.06 and 3.98.

21 Round 7296 to the nearest hundred.

22 The sum of a number and 93 is 116. What is the number?

1 $20\,000 + 50\,000 + 30\,000 =$ ☐

2 $968\,421 - 398\,689 =$ ☐

3 Find the product of 462 and 9.

4 Share 972 stickers equally between 6 children. How many stickers does each child receive?

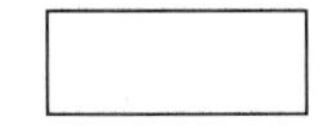

5 Circle the larger shaded fraction.

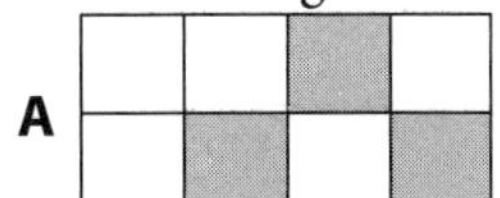

6 Use the diagram to find:

$\frac{5}{6} - \frac{3}{6} =$ ☐

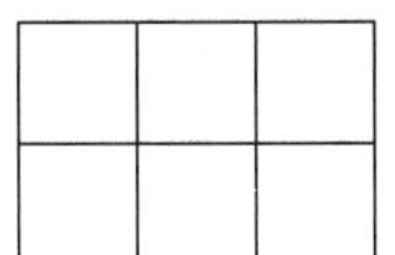

7 Stan has 50 counters. Jack has 35 more counters than Stan and Tyler has 17 more counters than Stan. How many counters are there altogether?

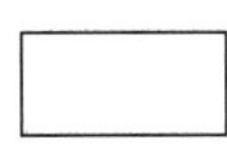

8 Draw 7:10 on the clock.

9 The movie starts at 2:10 pm and ends at 4:00 pm. How long is the movie?

10 How many metres is 5 km?

☐ m

11 Complete the table.

Length	Width	Perimeter
25 cm	30 cm	

12 Tick to select the most appropriate unit to measure the mass.

Object	g	kg	t
mass of a calculator			

13 Draw the shape rotated a three-quarter turn anticlockwise.

14 Name the shape of each of the views.

top view __________

front view __________

side view __________

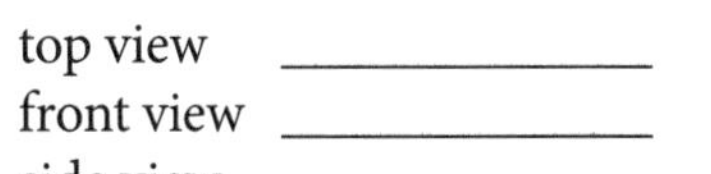

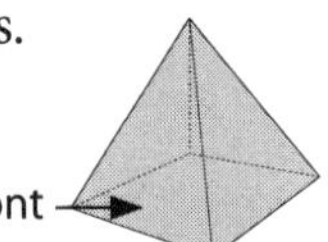

15 What is the name of the park to the south-west of Alice Springs?

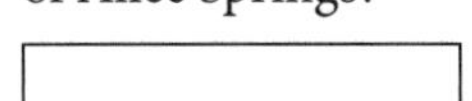

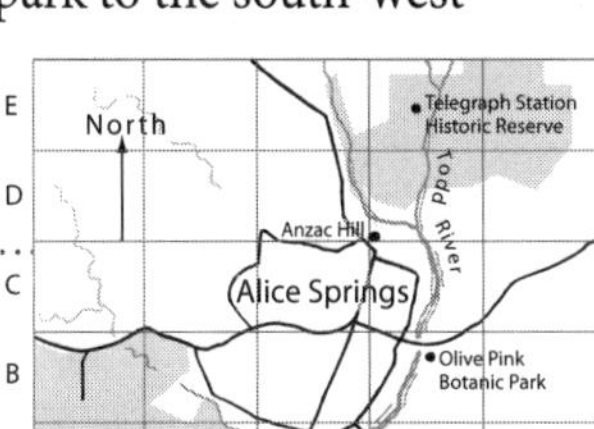

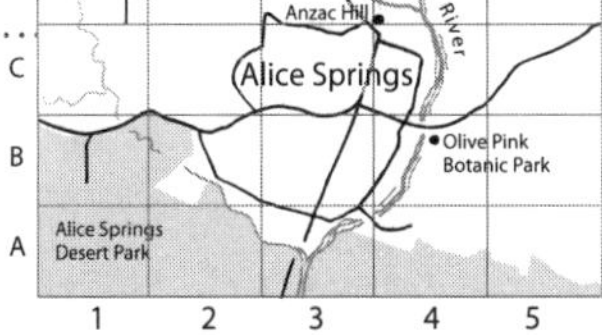

16 What is at 5C?

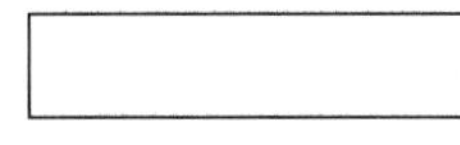

17 Circle the angle type for 260°.

straight   obtuse   acute   reflex   right

18 Which fruit has a 4 in 10 chance of being selected?

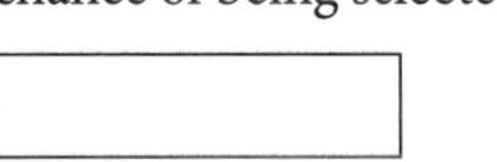

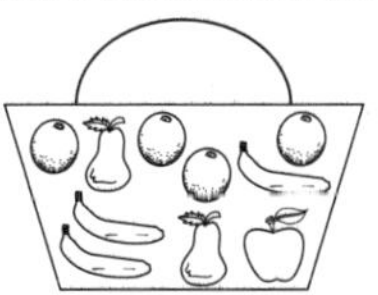

19 Mark the word **impossible** on the scale.

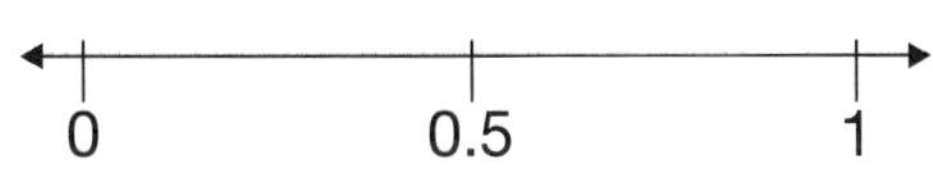

20 How many pieces of A3 paper did Bob own?

Paper	Number
A3	100
A4	200
A5	160

21 What was the total number of newspapers bought for the 5 days?

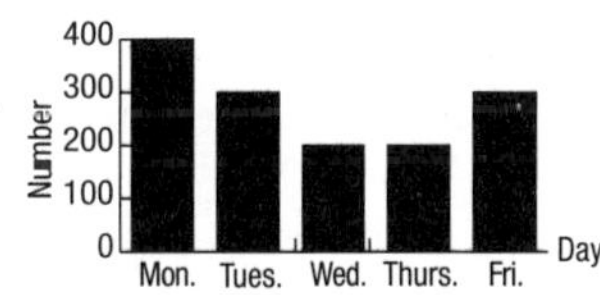

22 Create a dot plot for the following data.

No. of cars in household	No. of people surveyed
0	6
1	8
2	20
3	9

1

+	27	11	43	20	18	9
50						

2

–	94	89	76	50	67	81
30						

3

×	2	6	7	12	4	11
4						

4

×	1	5	8	9	11	3
10						

5

÷	70	28	84	35	14	77
7						

6

÷	60	90	30	10	110	40
10						

7

$$\begin{array}{r} 2476 \\ +\ 3917 \\ \hline \end{array}$$

8 What is the total of 412 L, 639 L and 285 L?

9 What is the difference between $9076 and $2005?

10 There are 1438 counters at a shop. 274 are sold. How many counters are left?

11

$6 \times 30 =$ ☐

$9 \times 80 =$ ☐

12

$$\begin{array}{r} 1172 \\ \times\ \ 4 \\ \hline \end{array}$$

13 Find the missing digits.

$$\begin{array}{r} 201 \\ 3\overline{)\square\square\square} \end{array}$$

14

$40 \div 6 =$ ☐

15 Circle the smaller number in each pair.

246 381 264 381

390 517 380 571

16 Order the numbers from smallest to largest.

2 082 463 2 083 498 2 085 775 2 085 762

17 How many pens can Victoria buy with $5.00 if one pen costs $1.05?

18 The movie treats cost $9.50. What is Joe's change from $10.00?

19 Order the decimals from smallest to largest.

0.19 0.2 0.11 0.5

20 Add 4.59 and 8.76.

21 Round 3009 to the nearest hundred.

22 The difference between a number and 85 is 17. What is the number?

1 $60\,000 + 70\,000 + 80\,000 =$ ☐

2 $369\,847 - 199\,608 =$ ☐

3 Find the product of 621 and 8.

☐

4 Share 684 cows equally between 4 fields. How many cows will be in each field?

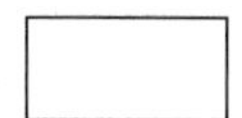

5 Circle the larger shaded fraction.

A 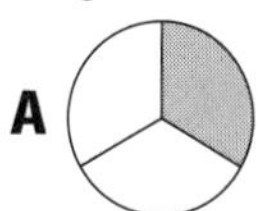 B 

6 Use the diagram to find

$\frac{9}{10} - \frac{7}{10} =$ ☐

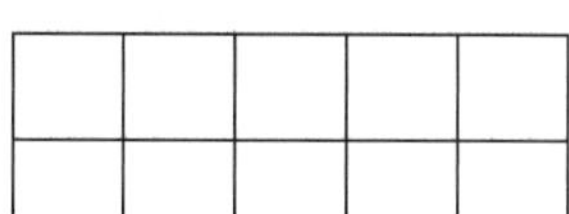

7 There are 60 pencils in each of 4 packets and there are 20 pencils in each of 9 packets. How many pencils are there altogether?

☐

8 Draw 11:25 on the clock.

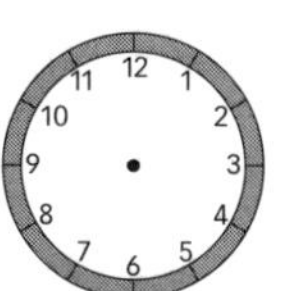

9 The concert starts at 11:00 am and finishes at 1:30 pm. How long is the concert?

☐

10 How many metres is 9 km?

☐ m

11 Complete the table.

Length	Width	Perimeter
17 cm	35 cm	

12 Tick to select the most appropriate unit to measure the mass.

Object	g	kg	t
mass of a submarine			

13 Draw the shape rotated a quarter turn anticlockwise.

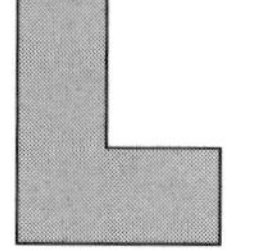

14 Name the shape of each of the views.

top view __________

front view __________

side view __________

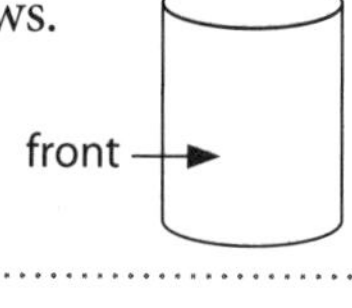

15 What is the name of the reserve to the north of Alice Springs?

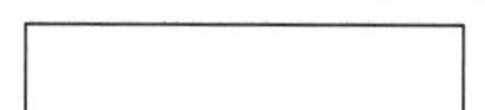

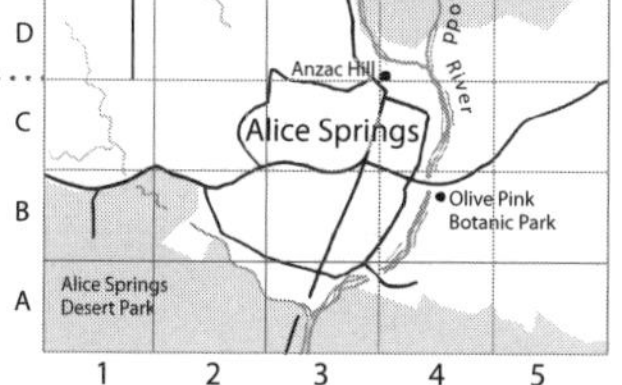

16 What is at 4B?

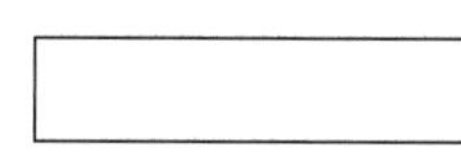

17 Circle the angle type for 320°.

straight obtuse acute reflex right

18 What is the chance of selecting a banana?

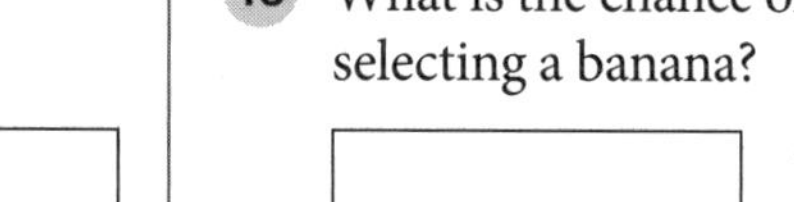

19 Mark the word **certain** on the scale.

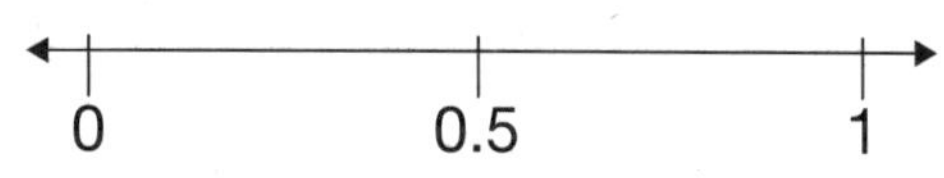

20 How many loaves of bread were in Pauline's shop? ☐

Food item	Number
loaves of bread	60
cartons of eggs	40
packets of bacon	10

21 What was the favourite activity of the most students? ☐

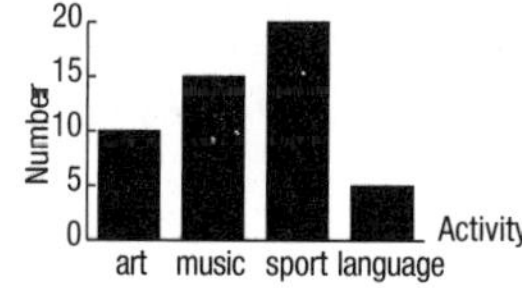

22 Create a dot plot for the following data.

No. of phones	No. of people
1	6
2	20
3	17
4	8

1

+	46	85	92	38	40	79
100						

2

–	250	385	462	396	201	450
200						

3

×	2	6	3	8	7	5
10						

4

×	1	9	10	11	4	12
3						

5

÷	110	40	90	60	10	100
10						

6

÷	44	16	8	36	12	40
4						

7 32 256 + 4199 = ☐

8 Add 13 000 and 17 000.

9 I have 486. How much more is needed to make 1700?

10 Alice has $1400. She spends $673. How much money does Alice have left?

11 There are 12 eggs in a carton. Sophia has 9 cartons. How many eggs does Sophia have?

12 What are the first 5 multiples of 3?

13 What are the factors of 20?

14 Lily had 90 pencils. She shared them equally between 9 pencil cases. How many pencils were in each pencil case?

15 Write the largest number possible using the digits 2, 1, 4, 3, 8, 3 and 0.

16 Write 204 638 in words.

17 Jack bought this ball. What was Jack's change from $20.00?

$11.50

18 Round $2.43 to the nearest 5c.

19 What is the difference between 3.8 and 2.96?

20 4.3 × 10 = ☐

21 Round each number to the nearest 10 to estimate the total.

326 + 409 + 217= ☐

22 Complete:

+		16		21
40	80		96	

1 Add 769 321 and 421 076.

2

$$\begin{array}{r} 236\,789 \\ -\ 147\,325 \\ \hline \end{array}$$

3 Find the total of 9 groups of 125.

4

$$6452 \div 4 = \square$$

5

$$1 - \frac{1}{4} = \square$$

6

$$\frac{2}{3} + \frac{1}{3} = \square$$

7 Complete the brackets first.

$$60 + (4 \times 10) - 19 = \square$$

8 Match the times.

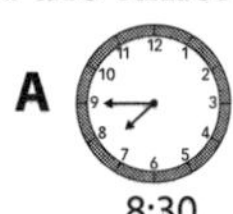

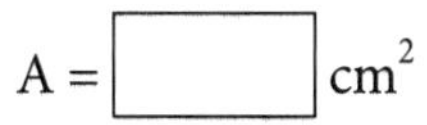

A B C

8:30 8:15 7:45

9 Circle the earliest time.

1:15 am 1:08 am 1:30 pm

10 Rewrite 2750 g as kilograms and grams.

_______ kg _______ g

11 Find the area.

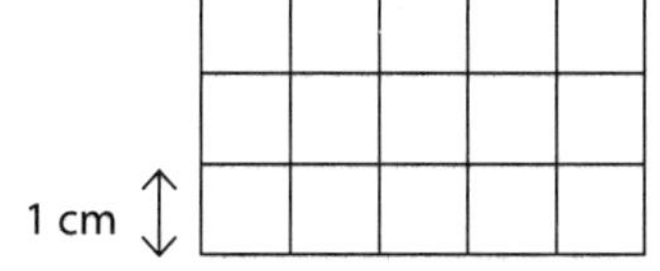

A = $\square$ cm^2

12 Complete the table for the model.

Length (cm)	Width (cm)	Height (cm)	Volume (cm^3)

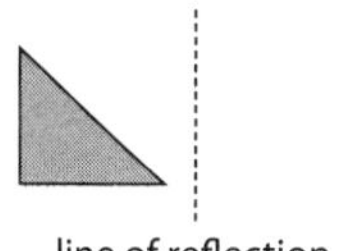

13 Draw the reflection of the triangle.

14 Match the net with the object.

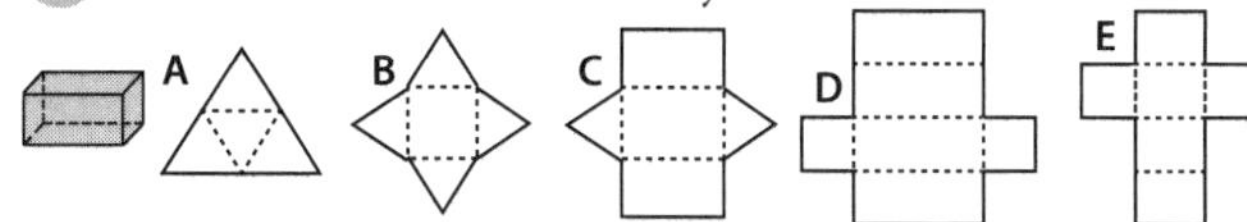

15 Label with the name **Rose** the book that is 3 away from Will's book.

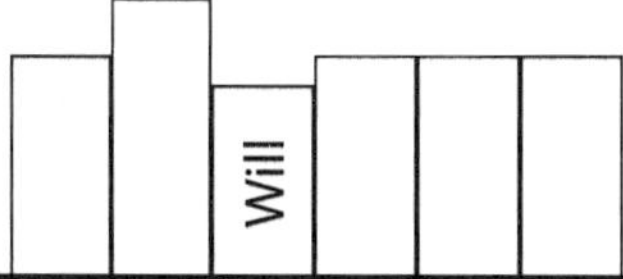

16 Draw a:

- star at (3, 4)
- diamond at (6, 2)
- circle at (4, 1)

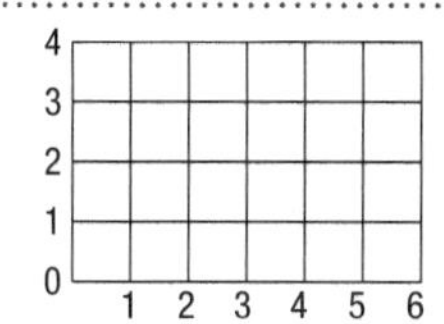

17 What is the size of the shaded angle?

18 Mark the scale to show the chance of landing on yellow.

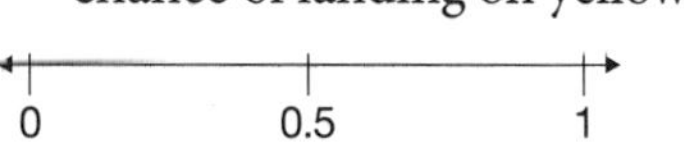

19 List all of the outcomes for this spinner.

20 Complete the table to show the data.

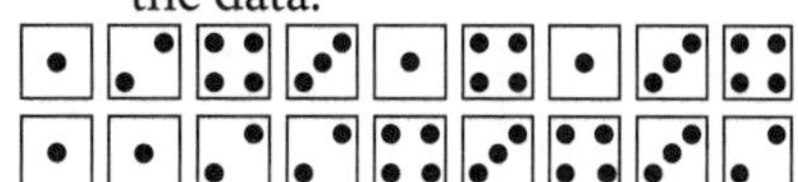
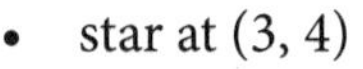
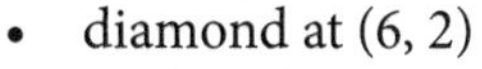
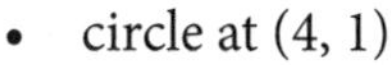

Dice number	Number
1	
2	
3	
4	

21 Draw a column graph for the data.

Colour	Number
red	15
gold	16
black	22
silver	29

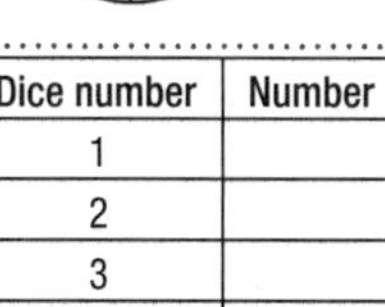

22 Which berry has a value of 15?

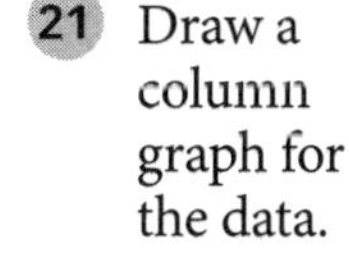

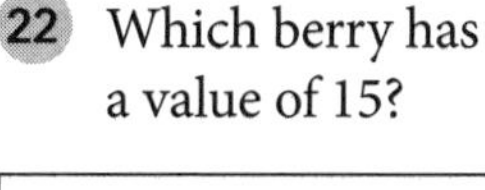

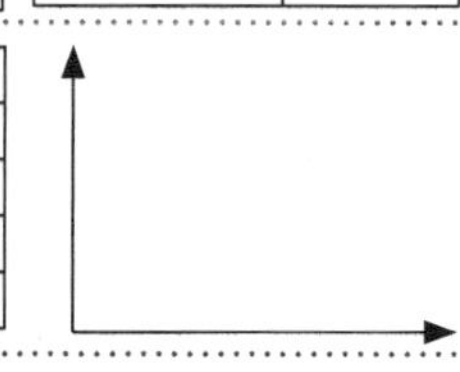

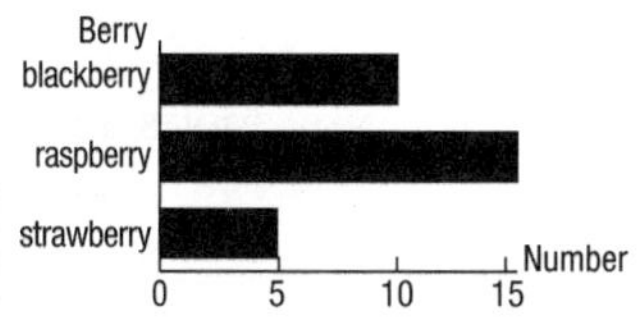

1

+	326	144	38	99	102	439
200						

2

–	463	246	145	287	311	107
100						

3

×	10	6	11	2	7	9
10						

4

×	4	12	3	8	5	1
5						

5

÷	20	90	30	50	80	10
10						

6

÷	90	36	72	45	81	27
9						

7 76 729 + 30 149 = ☐

8 Add 14 000 and 39 000.

9 I have 346. How much more is needed to make 1800?

10 Amity has 197 m of ribbon at the fabric store. She sells 29 m. How much ribbon does Amity have left?

11 Apples come in bags of 8. George has 28 bags. How many apples does George have?

12 What are the first 5 multiples of 4?

13 What are the factors of 40?

14 Zen had 50 lemons. He shared them equally between 10 bowls. How many lemons were in each bowl?

15 Write the largest number possible using the digits 6, 2, 1, 0, 3, 9 and 2.

16 Write 2 170 850 in words.

17 Jill bought this book. What was Jill's change from $20.00?

$15.95

18 Round $6.77 to the nearest 5c.

19 What is the difference between 4.6 and 3.25?

20 $7.9 \times 10 =$ ☐

21 Round each number to the nearest 10 to estimate the total.

407 + 392 + 176 = ☐

22 Complete:

+	19		26	
17		35		50

1 Add 673 247 and 480 329.

2

$$\begin{array}{r} 463\,117 \\ -\ 269\,085 \\ \hline \end{array}$$

3 Find the total of 7 groups of 367.

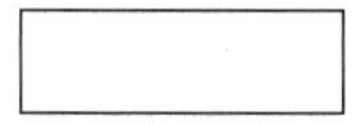

4

$$6456 \div 8 = \square$$

5

$$1 - \frac{1}{6} = \square$$

6

$$\frac{2}{5} + \frac{1}{5} = \square$$

7 Complete the brackets first.

$$150 + (5 \times 20) - 72 = \square$$

8 Match the times.

A  B  C 

1:30 12:45 2:15

9 Circle the earliest time.

10:50 pm 11:00 pm 9:00 pm

10 Rewrite 1160 g as kilograms and grams.

_______ kg _______ g

11 Find the area.

A = ☐ cm^2

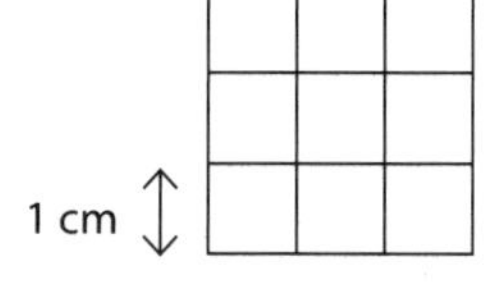

12 Complete the table for the model.

Length (cm)	Width (cm)	Height (cm)	Volume (cm^3)

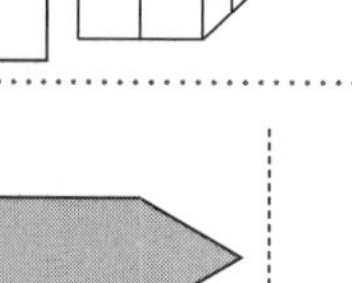

13 Draw the reflection of the shape.

line of reflection

14 Match the net with the object.

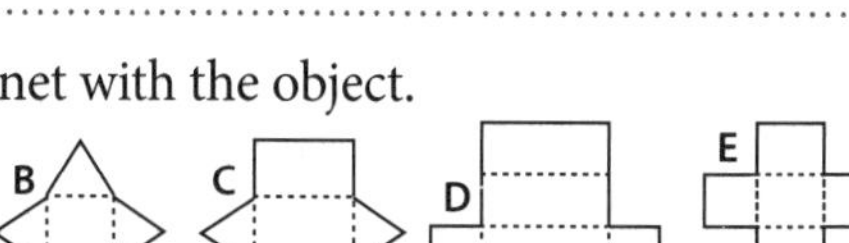

15 Label with the name **Penny** the book that is on the right next to Will's book.

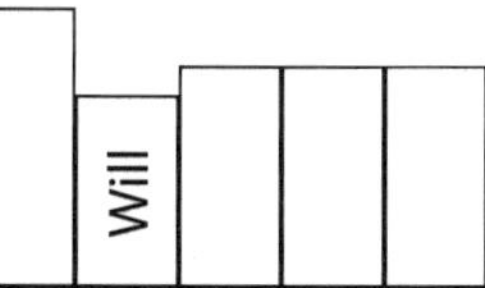

16 Draw a:

- smiley face at (3, 2)
- flower at (6, 4)
- tennis racquet at (3, 0)

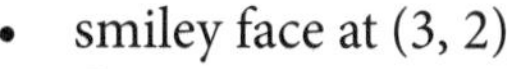

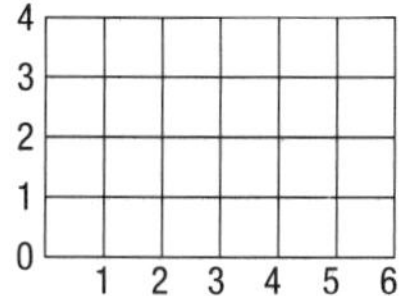

17 What is the size of the shaded angle?

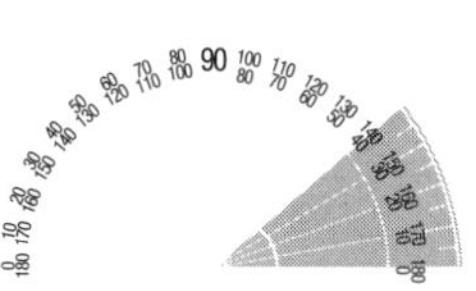

18 Mark the scale to show the chance of landing on green.

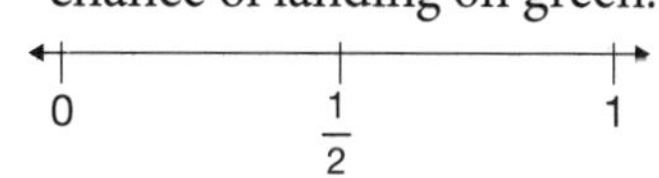

19 List all of the outcomes for this spinner.

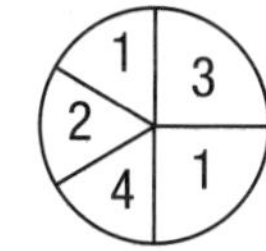

20 Complete the table to show the data.

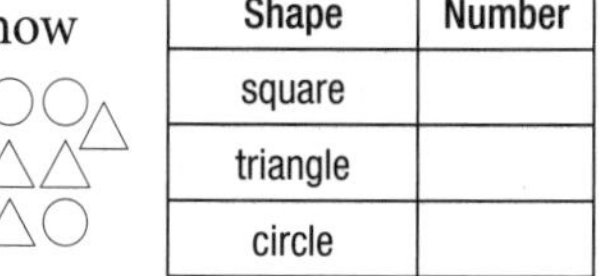

Shape	Number
square	
triangle	
circle	

21 Draw a column graph for the data.

Technology	Number
laptop	50
PC	32
tablet	65
smartphone	72

22 Which month has the greatest number of tourists?

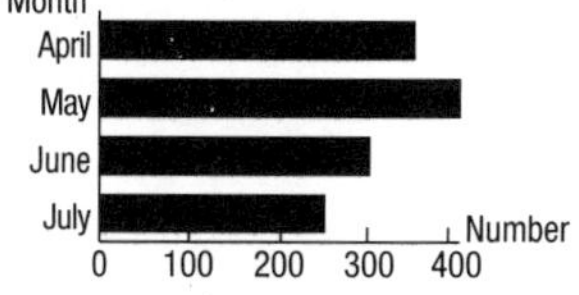

1

+	271	119	485	308	25	199
300						

2

−	385	463	102	573	219	350
100						

3

×	7	6	3	8	2	9
10						

4

×	4	10	11	5	12	0
7						

5

÷	40	60	20	90	80	30
10						

6

÷	80	32	8	24	56	48
8						

7 $67\,849 + 21\,736 = \square$

8 Add 17 000 and 34 000.

9 I have 9243. How much more is needed to make 10 500?

10 Alexander has 327 football cards. He gives 185 cards away. How many football cards does Alexander have left?

11 There are 12 pencils in a box. The teacher has 9 boxes. How many pencils does the teacher have?

12 What are the first 5 multiples of 6?

13 What are the factors of 24?

14 Sam had 80 peaches. He put 8 peaches in each box. How many boxes did Sam have?

15 Write the largest number possible using the digits 6, 2, 1, 1, 7, 9 and 3.

16 Write 2 016 209 in words.

17 Hazel bought this dog coat. What was Hazel's change from $20.00?

$12.50

Rusty

18 Round $1.22 to the nearest 5c.

19 What is the difference between 4.06 and 9.15?

20 $8.6 \times 10 = \square$

21 Round each number to the nearest 10 to estimate the total.

$407 + 304 + 627 = \square$

22 Complete:

+		17	55	
29	60			75

UNIT 23B

1 Add 576 042 and 321 796.

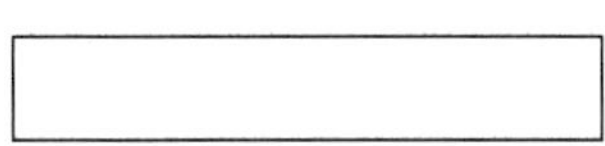

2

$$\begin{array}{r} 796\,043 \\ -\ 279\,867 \\ \hline \end{array}$$

3 Find the total of 8 groups of 279.

4

$6456 \div 6 =$ ☐

5

$1 - \frac{1}{8} =$ ☐

6

$\frac{3}{8} + \frac{2}{8} =$ ☐

7 Complete the brackets first.

$50 + (12 \times 9) - 66 =$ ☐

8 Match the times.

A B C

12:00 12:15 11:45

9 Circle the earliest time.

3:30 pm 3:45 pm 2:40 pm

10 Rewrite 4700 g as kilograms and grams.

_______ kg _______ g

11 Find the area.

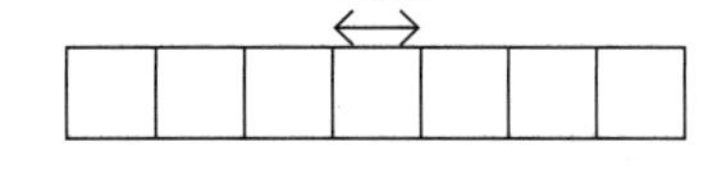

A = ☐ cm^2

12 Complete the table for the model.

Length (cm)	Width (cm)	Height (cm)	Volume (cm^3)

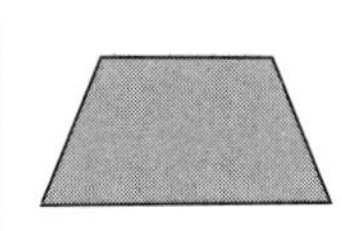

13 Draw the reflection of the shape.

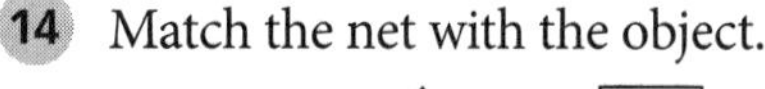
line of reflection

14 Match the net with the object.

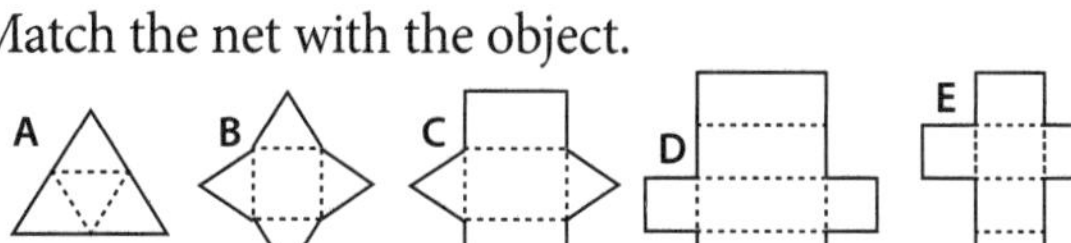

15 Label with the name **Melissa** the book that is the furthest to the left.

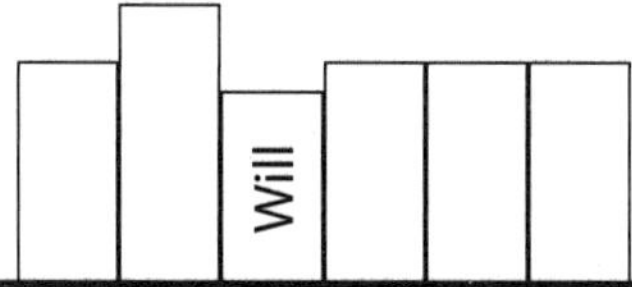

16 Draw a:
- sun at (0, 2)
- square at (5, 4)
- heart at (2,1)

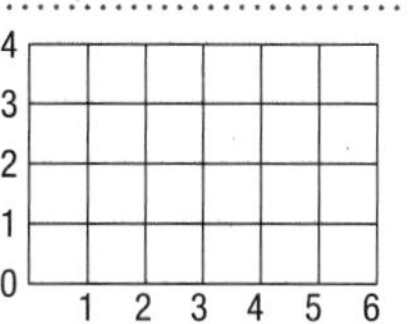

17 What is the size of the shaded angle?

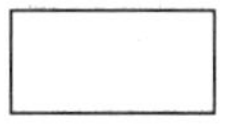
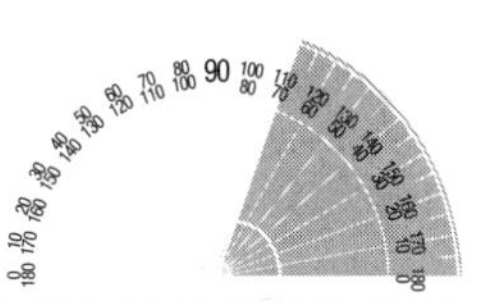

18 Mark the scale to show the chance of landing on green.

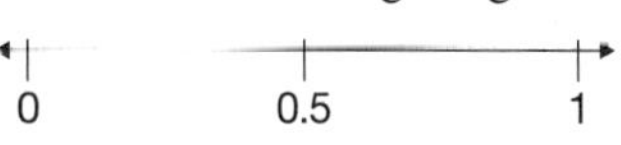

19 List all of the outcomes for this spinner.

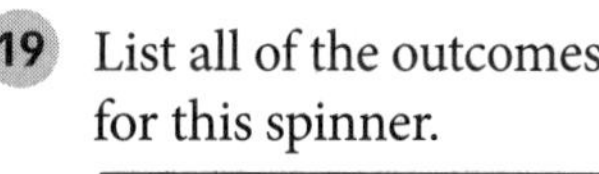

20 Complete the table to show the data.

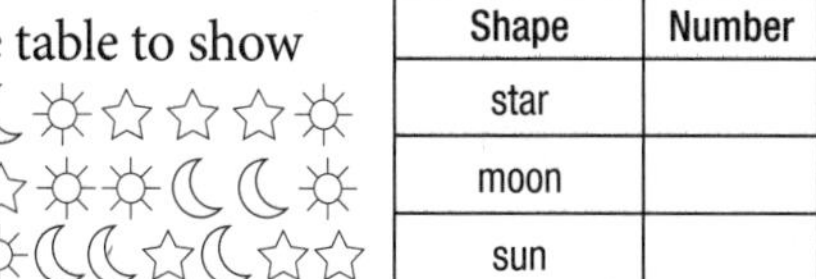

Shape	Number
star	
moon	
sun	

21 Draw a column graph for the data.

Day	Number
Monday	90
Tuesday	72
Wednesday	85
Thursday	102

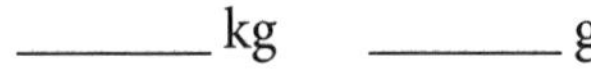
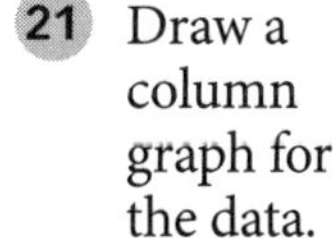
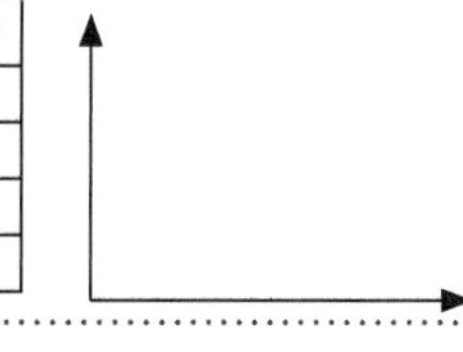

22 On which day was the greatest number of whales seen?

1

+	16	24	92	35	70	63
20						

2

–	108	97	62	50	79	88
40						

3

×	7	9	3	12	1	10
4						

4

×	2	5	8	4	11	6
8						

5

÷	77	28	49	56	63	14
7						

6

÷	30	9	18	15	6	36
3						

7 Joe has two cards. He added the numbers. What was the total?

4173

2789

A 6952 **B** 6962
C 6852 **D** 6862

8 Here are the lengths of three strips of paper. What is the total length?

96 mm
60 mm
73 mm

A 129 mm **B** 133 mm
C 229 mm **D** 169 mm

9 13 000 – 1700 = ☐

10 There were 1400 birds in the field. 985 flew away. How many birds were left?

☐

11 Complete with >, < or =.

90 × 4 ☐ 8 × 50

12

$$\begin{array}{r} 4798 \\ \times \quad 6 \\ \hline \end{array}$$

13 List all of the factors of 18.

☐

14 21 ÷ 6 = ☐

15 Which is the smallest number?

A 7 171 396 **B** 7 171 936
C 7 171 963 **D** 7 171 369

16 Write 1 765 200 in words.

☐

17 James bought this fruit salad. What was James's change from $20.00? ☐

$7.80

18 One sushi roll costs $2.30. How many sushi rolls can Mitchell buy for $10?

☐

19 What is the sum of 3.8 and 7.62?

☐

20 A bag of dog food weighs 4.3 kg. How much do 8 bags weigh?

☐

21 Complete the table.

Number	3216	40 793	118 306
Round to the nearest 10			
Round to the nearest 100			

22 The difference between a number and 27 is 52. What is the number?

☐

1

$40\,000 + 90\,000 + 115\,000 =$ ☐

2 How much larger than 118 392 is 392 561?

3 True or false?

487 × 6 is less than 2000.

4 924 golf balls are shared equally into 7 boxes. How many golf balls are in each box?

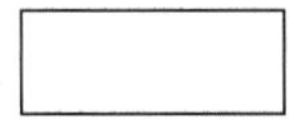

5 What is the largest fraction?

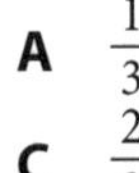

A $\frac{1}{3}$ **B** $\frac{5}{6}$

C $\frac{2}{3}$ **D** $\frac{1}{6}$

6

$\frac{2}{5} + \frac{1}{5} =$ ☐

7 There are two different types of packets of bread rolls. The shop has 10 packets of 8 rolls and 12 packets of 6 rolls. How many bread rolls are there altogether?

8 Which clock shows 7:45?

A

B

C

D

9 The movie starts at 11:45 am and finishes at 2:30 pm. How long is the movie?

10 Complete the table.

Mass	4250 g	3060 g	1490 g
kg	4		
g	250		

11 Find the area.

A = 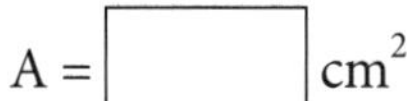 cm^2

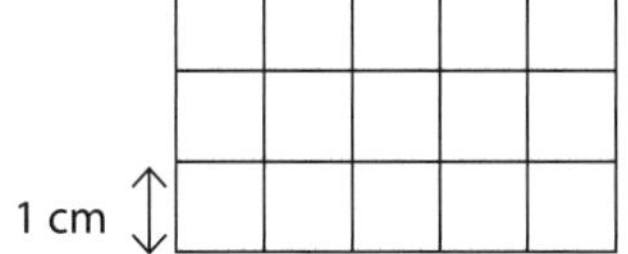

12 Complete the table for the model.

Length (cm)	Width (cm)	Height (cm)	Volume (cm^3)

13 Draw the reflection of the shape.

line of reflection

14 Draw the top view of the cone.

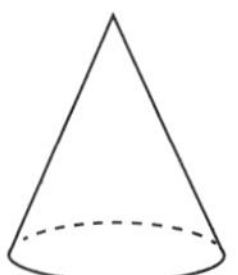

15 What is north-west of the person?

16 On the grid draw a cross at (2, 3) and a dot at (3, 0).

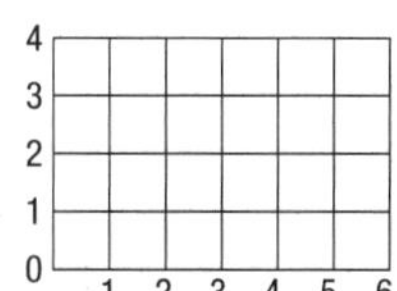

17 What is the size of the shaded angle?

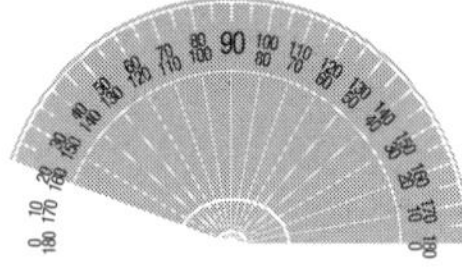

18 What is the chance of randomly selecting a book from the pile of reading materials?

19 Show on the scale the chance of rolling an even number on a standard six-sided dice.

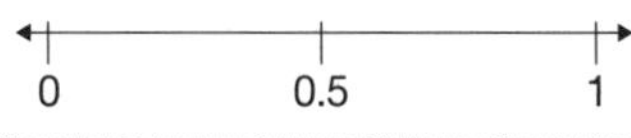

20 Use the graph to complete the table.

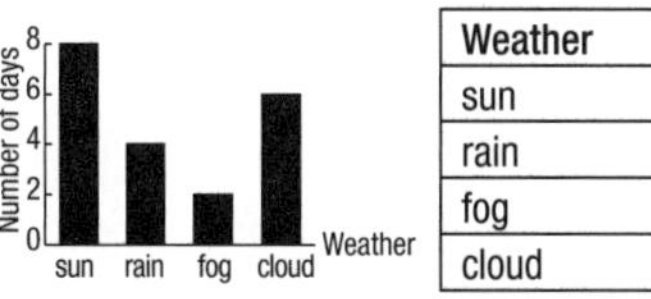

Weather	Number
sun	
rain	
fog	
cloud	

21 How many basketballs does Shop 3 have?

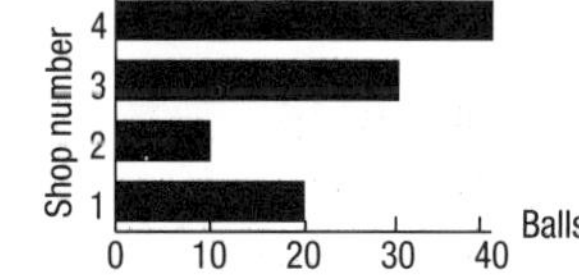

22 Draw a column graph for the data.

Time	Tally
9 am	90
10 am	72
11 am	85
12 noon	102

No. of people entering zoo

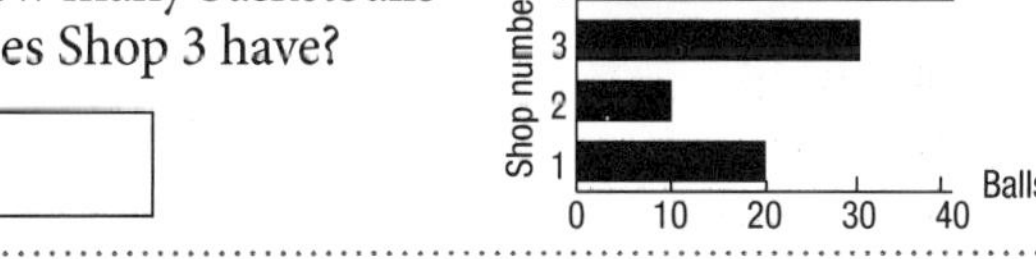

1 Arthur added three numbers 1462, 2135 and 3293. What was Arthur's answer?

A 5790
B 6880
C 6888
D 6890

2 There are 350 football cards in one set. Sally has 173 cards. How many more cards does Sally need to make the full set?

A 177
B 277
C 273
D 223

3 Which multiplication gives the largest value?

A 9×60
B 8×70
C 10×50
D 11×55

4 What is the remainder of $50 \div 9$?

A 1
B 4
C 5
D 6

5 What is ninety-nine thousand, three hundred and six written as a number?

A 9936
B 99 306
C 990 306
D 909 306

6 Which number is **not** a factor of 60?

A 4
B 6
C 8
D 10

7 Nadia bought these two items. What was Nadia's change from $10?

A $5.60
B $5.50
C $6.50
D $6.60

8 What is the difference between 4.87 and 6.09?

9 There are 1695 cars in one car park and 3467 cars in another.
How many cars are in the two car parks altogether?

A 5171
B 5062
C 5052
D 5162

10 Mum opened the box and saw there was $\frac{3}{8}$ of her cake left. How much of the cake had been eaten?

A $\frac{3}{8}$ **B** $\frac{5}{8}$ **C** $\frac{6}{8}$ **D** $\frac{8}{8}$

11 For every 3 bottles of water purchased the customer receives 2 bananas for free. Jack bought 15 bottles of water. How many bananas did he get for free?

A 5
B 6
C 10
D 20

12 Complete the equation with <, > or =.

$$9 \times 6 - 20 \; \square \; 70 \div 2 + 3$$

13 Which clock shows $\frac{1}{2}$ hour later than 3:45?

A

B

C

D

14 How many grams is 3 kg?

A 30 g
B 300 g
C 3000 g
D 30 000 g

15 Find the area of the shaded shape.

A 5 cm^2
B 6 cm^2
C 10 cm^2
D 25 cm^2

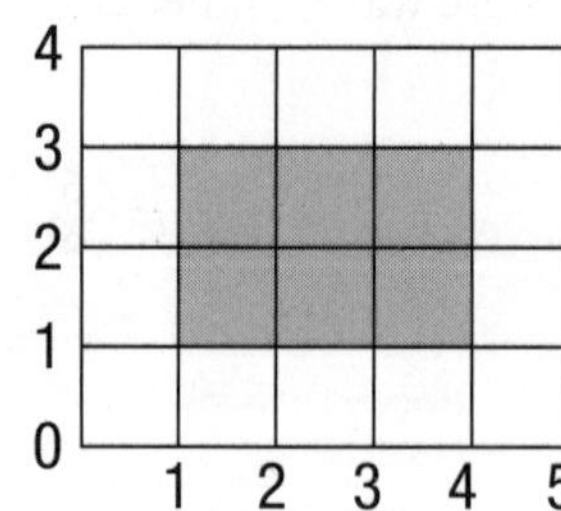

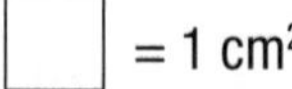

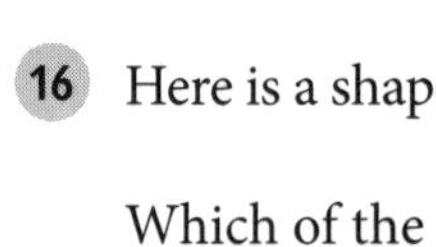

16 Here is a shape.

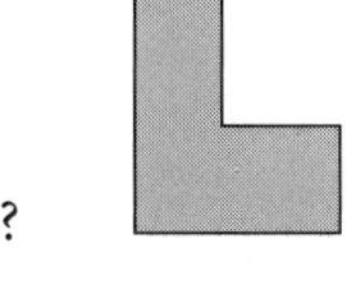

Which of the following is the shape rotated a quarter turn anticlockwise?

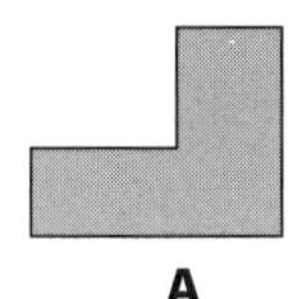

A

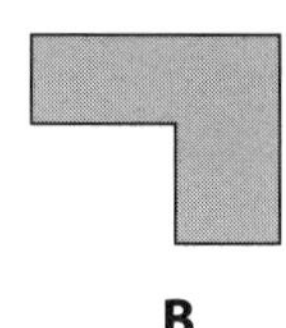

B

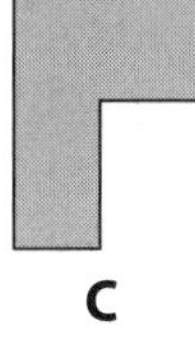

C

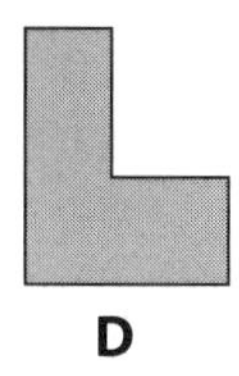

D

---

17 What is the size of the shaded angle?

A 30°
B 90°
C 70°
D 110°

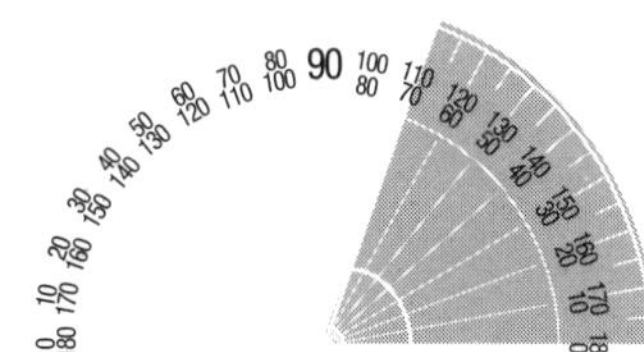

---

18 What is the chance of getting two heads when throwing two coins?

A 1 in 4
B 1 in 6
C 1 in 2
D 1 in 8

---

19 Here is a graph of the number of bottles of orange juice sold over three days.
What is the total number of bottles sold?

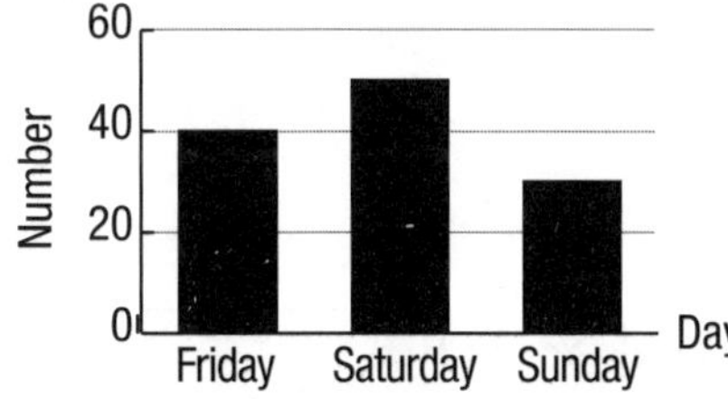

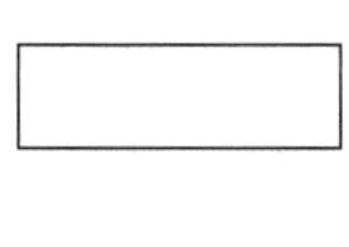

---

20 Different pieces of wood are measured and recorded on the graph.
Piece number 4 is 3 m longer than piece number 2.
What is the length of piece number 4?

A 2.5 m
B 5.5 m
C 6 m
D 7 m

Piece 4
Piece 3
Piece 2
Piece 1
0 1 2 3 4 5 Length m

UNIT 24A

1

+	218	466	110	325	172	248
400						

2

–	211	360	420	487	279	308
200						

3

×	10	5	8	7	2	11
10						

4

×	3	9	4	12	6	0
6						

5

÷	100	60	70	40	10	110
10						

6

÷	56	77	21	35	14	49
7						

7 $32\,379 + 7841 =$ ☐

8 Add 16 000 and 19 000.

9 I have 1318. How much more is needed to make 6000?

10 Emily is packaging 1486 buttons. She puts 790 into containers. How many buttons does Emily have left to package?

11 There are 6 tennis balls in a packet. The sports teacher has 25 packets. How many tennis balls does the sports teacher have?

12 What are the first 5 multiples of 7?

13 What are the factors of 30?

14 Mel had 60 flowers. She placed them equally in 5 vases. How many flowers were in each vase?

15 Write the largest number possible using the digits 2, 9, 3, 8, 4, 6 and 1.

16 Write 6 042 179 in words.

17 Phillip bought some pet food. What was Phillip's change from $20.00?

18 Round $3.66 to the nearest 5c.

19 What is the difference between 7.72 and 9.35?

20 $6.4 \times 10 =$ ☐

21 Round each number to the nearest 10 to estimate the total.

$356 + 469 + 207 =$ ☐

22 Complete:

+	17			28
35		49	87	

1 Add 796 321 and 463 427.

2

$$\begin{array}{r} 530\,695 \\ -\ 243\,046 \\ \hline \end{array}$$

3 Find the total of 7 groups of 462.

4

$$6452 \div 4 = \square$$

5

$$1 - \frac{1}{10} = \square$$

6

$$\frac{4}{10} + \frac{5}{10} = \square$$

7 Complete the brackets first.

$$100 - (9 \times 8) + 15 = \square$$

8 Match the times.

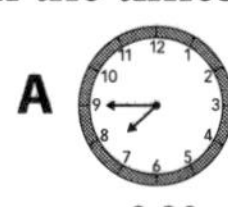

A B C

8:00 7:45 8:30

9 Circle the earliest time.

2:05 am 1:50 am 1:30 pm

10 Rewrite 3950 g as kilograms and grams.

_______ kg _______ g

11 Find the area.

A = $\square$ cm^2

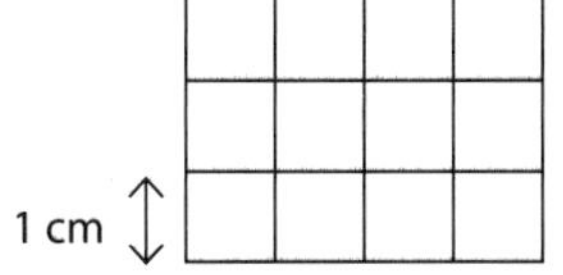

12 Complete the table for the model.

Length (cm)	Width (cm)	Height (cm)	Volume (cm^3)

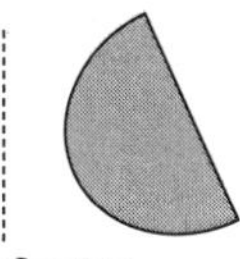

13 Draw the reflection of the shape.

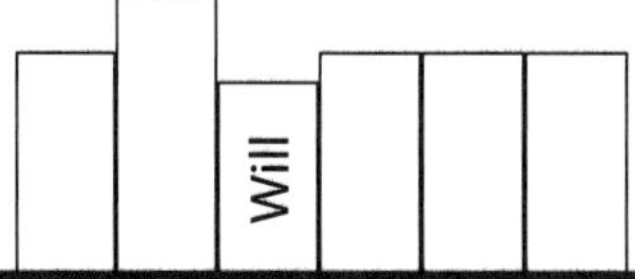

14 Match the net with the object.

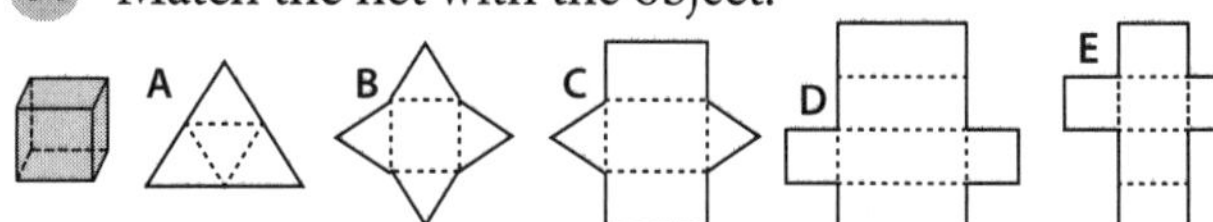

15 Label with the name **Jack** the book that is on the right next to Will's.

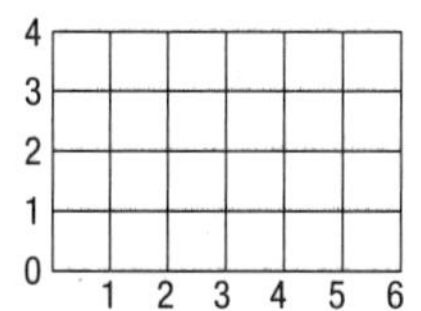

16 Draw a:

- triangle at (5, 0)
- star at (2, 4)
- worm at (4, 3)

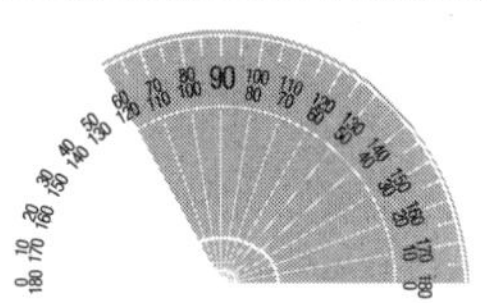

17 Measure the size of the shaded angle.

18 Mark the scale to show the chance of landing on white.

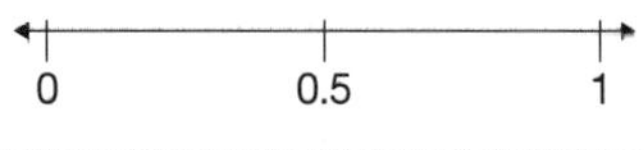

19 List all of the outcomes for this spinner.

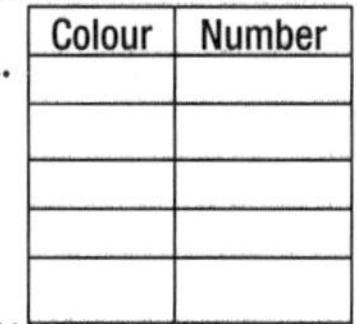

20 Complete the table to show the data.

red, green, blue, yellow, white, red, green, red, blue, yellow, white, yellow, red, white, red, red, green, green, white, blue, red, red, green, red, yellow, white, green, green, blue, blue, white, yellow, yellow, white

Colour	Number

21 Draw a column graph for the data.

Note	Number
$50	16
$20	95
$10	70
$5	53

22 How many people have green eyes?

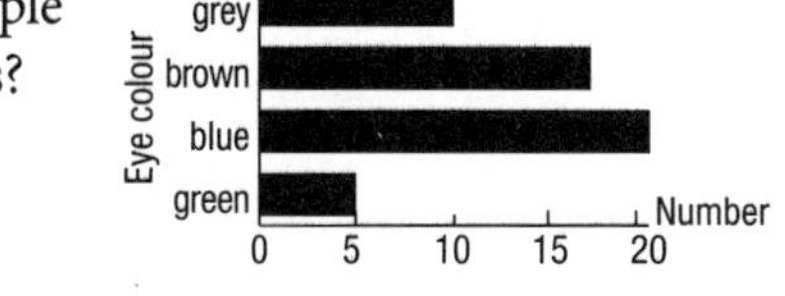

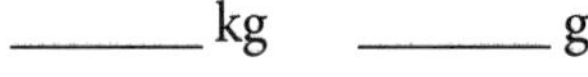

UNIT 25A

1

+	423	109	238	349	287	177
100						

2

–	201	627	435	855	580	463
100						

3

×	6	3	9	1	8	7
10						

4

×	4	10	2	11	12	5
8						

5

÷	120	10	60	50	20	90
10						

6

÷	24	30	6	18	9	27
3						

7 32 249 + 17 763 = ☐

8 Add 1500 and 2800.

9 I have 7703. How much more is needed to make 10 000?

10 Brian has 1850 flowers to put into bunches to sell. He uses 685. How many flowers does he left for bunches?

11 There are 8 boxes with 14 chocolates in each. How many chocolates are there altogether?

12 What are the first 5 multiples of 3?

13 What are the factors of 18?

14 Max had 45 balls. He placed 5 balls in each bucket. How many buckets did Max have?

15 Write the largest number possible using the digits 4, 9, 3, 8, 5, 1 and 7.

16 Write 6180 321 in words.

17 Billie bought this T-shirt. What was Billie's change from $20.00?

$8.25

18 Round $6.11 to the nearest 5c.

19 What is the difference between 3.06 and 7.28?

20 $8.6 \times 10 =$ ☐

21 Round each number to the nearest 10 to estimate the total.

614 + 289 + 301 = ☐

22 Complete:

+	46		28	
39		48		76

1 Add 273 986 and 427 106.

2

$$\begin{array}{r} 962\,000 \\ -\ 176\,432 \\ \hline \end{array}$$

3 Find the total of 6 groups of 485.

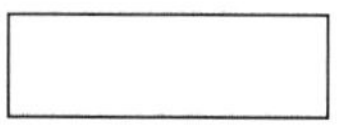

4

$6936 \div 3 = \square$

5

$1 - \frac{1}{3} = \square$

6

$\frac{1}{6} + \frac{4}{6} = \square$

7 Complete the brackets first.

$200 - (11 \times 6) + 59 = \square$

8 Match the times.

A  B C 

10:15 10:00 10:30

9 Circle the earliest time.

4:15 am 4:50 am 4:07 pm

10 Rewrite 5900 g as kilograms and grams.

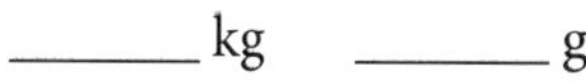

_______ kg _______ g

11 Find the area.

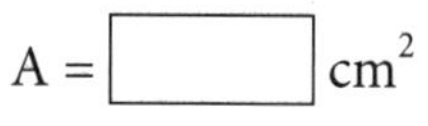

A = $\square$ cm²

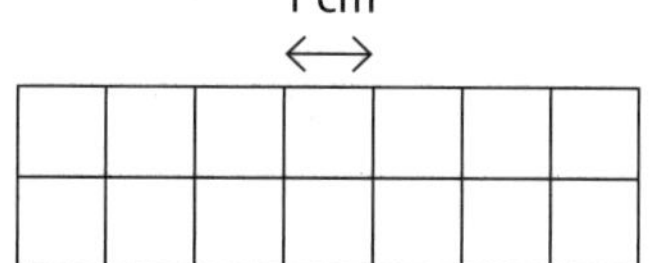

12 Complete the table for the model.

Length (cm)	Width (cm)	Height (cm)	Volume (cm³)

1 cm

1 cm

13 Draw the reflection of the shape.

line of reflection

14 Match the net with the object.

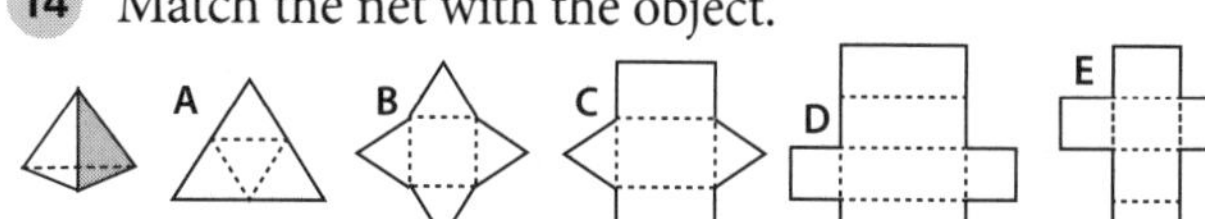

15 Label the fifth book from the right with the name **Zelda**.

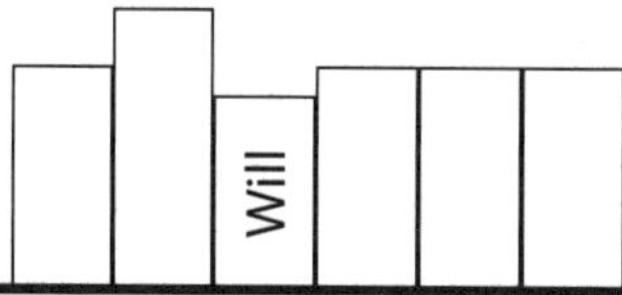

16 Draw a:

- cross at (6, 0)
- circle at (0, 4)
- tick at (3, 3)

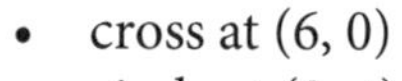

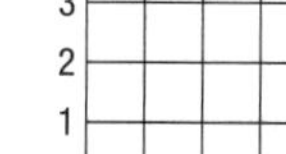

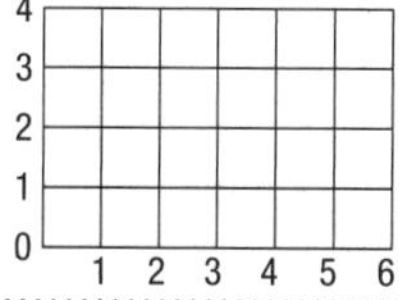

17 Measure the size of the shaded angle.

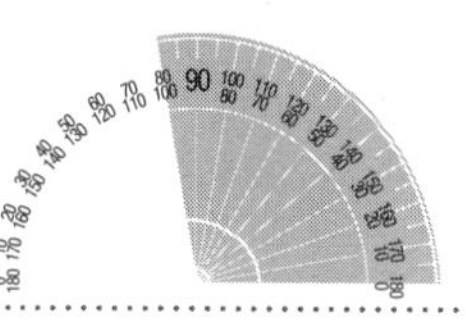

18 Mark the scale to show the chance of landing on white.

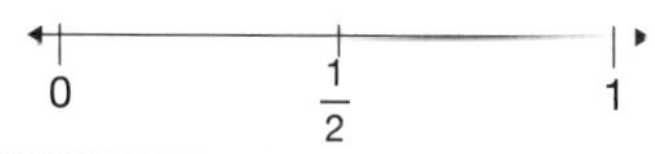

19 List all of the outcomes for this spinner.

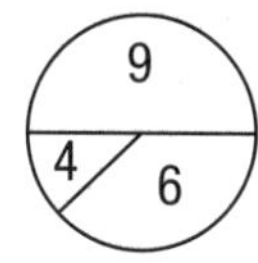

20 Complete the table to show the data.

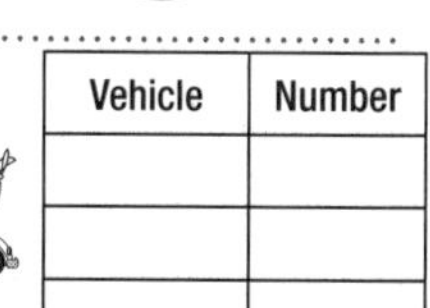

Vehicle	Number

21 Draw a column graph for the data.

Clothing	Number
T-shirt	12
shorts	8
dress	2
jumper	5

22 How many sheep and goats does the farmer have?

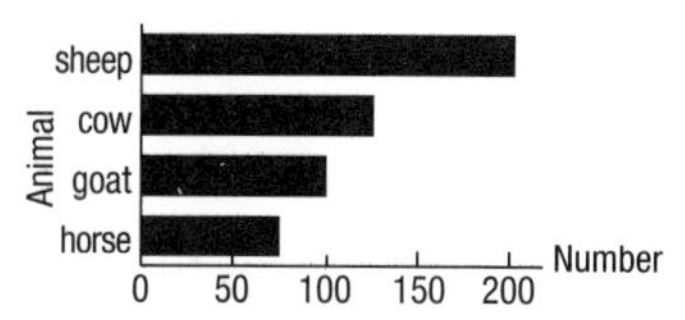

1

+	28	36	47	17	98	85
17						

2

–	27	46	85	72	91	50
19						

3

×	4	7	8	9	6	3
100						

4

×	1	11	5	10	12	2
8						

5

÷	800	100	700	200	400	900
100						

6

÷	49	35	63	21	56	70
7						

7 Complete the missing numbers.

```
    3 2 □ □
+   1 5 2 6
-----------
    □ □ 8 2
```

8 What is the total of 4000, 632 and 1484?

9

```
  64 000
– 27 809
```

10 What is 1400 – 120 – 35?

11 What are the first 5 multiples of 8?

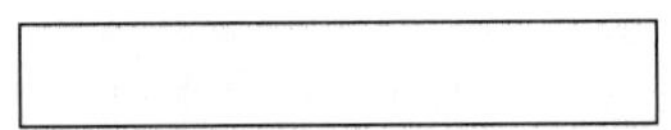

12 What is 6^2?

13 $925 \div 10 = \square$

14 What are the factors of 16?

15 What is the number of thousands in 143 286?

16 Insert <, > or =.

40 000 + 200 + 1 □ 42 000 + 20 + 1

17 Complete and round the answer to the nearest 5c.

```
  $13.93
+  $4.60
```

18 Complete the table.

Starting amount	Spent	Change
$50.00	$25.60	

19 $3 - 1.65 = \square$

20 Round 3.26 to the nearest tenth.

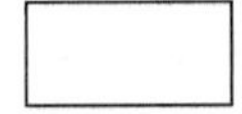

21 Round $13.65 to the nearest dollar.

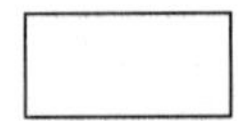

22 $136 - 47 = 18 + \square$

1

$$\begin{array}{r} 738\,941 \\ +\ 276\,847 \\ \hline \end{array}$$

2 What is the difference between 210 000 and 80 000?

3 How many days are there in 26 weeks?

4

$$4\overline{)4784}$$

5 Circle the smallest fraction.

$\frac{1}{3}$ $\frac{1}{4}$ $\frac{1}{8}$

6

$\frac{2}{5} + \frac{2}{5} = \square$

7 Start at 50, divide by 5, add 7 and multiply by 6. What is the answer?

8 What time is 25 minutes after 11:30 am?

9 Draw **2 : 10** on the clock.

10 Circle the longest.

120 cm 1 m 150 mm

11 Find the area.

A = $\square$ cm^2

2 cm

4 cm

12 Write 289 g to the nearest 10 g.

13 How many axes of symmetry does the shape have?

14 Name the shape.

15 Draw a ship to the north of the island.

16 What is the distance, in units, between C1 and C3?

17 Record the size of the shaded angle.

18 Give an example of an event that has a 0.5 chance of occurring.

19 What is the chance of selecting a stripey fish?

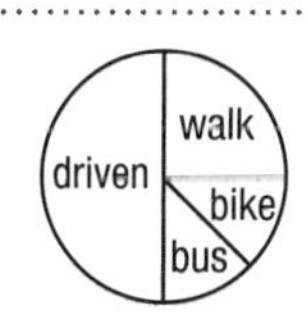

20 Complete the table of data using the graph.

Colour	Number
gold	
silver	
bronze	

21 What fraction of students are driven to school?

22 Draw a horizontal column graph for the data.

Shape	Number
circle	7
triangle	9
diamond	10
square	15

1

+	35	40	87	69	21	17
22						

2

–	20	65	36	73	41	85
11						

3

×	11	9	2	3	4	7
10						

4

×	1	6	10	8	12	5
6						

5

÷	500	600	800	200	300	1000
100						

6

÷	24	0	12	36	4	16
4						

7 Complete the missing numbers.

```
    4 6 3 8
  + □ 1 2 □
  ---------
    7 □ □ 9
```

8 What is the total of 4850, 329 and 647?

☐

9

```
   85 000
 – 37 263
```

10 What is 9000 – 1325 – 642?

☐

11 What are the first 5 multiples of 6?

☐

12 What is 8^2?

☐

13

$645 \div 10 = \square$

14 What are the factors of 50?

☐

15 What is the number of thousands in 217 031?

☐

16 Insert <, > or =.

2000 + 300 + 6 ☐ 20 000 + 30 + 6

17 Complete and round the answer to the nearest 5c.

```
   $20.00
 – $14.38
```

☐

18 Complete the table.

Starting amount	Spent	Change
$50.00	$13.70	

19

$14 - 6.72 = \square$

20 Round 4.06 to the nearest tenth.

☐

21 Round $14.06 to the nearest dollar.

☐

22

$240 - 183 = 21 + \square$

1

$$\begin{array}{r} 694\,328 \\ +\ 107\,463 \\ \hline \end{array}$$

2 What is the difference between 400 000 and 90 000?

3 How many hours are there in 9 days?

4

$$7\overline{)7028}$$

5 Circle the smallest fraction.

$\frac{1}{10}$ $\frac{1}{8}$ $\frac{1}{6}$

6

$\frac{7}{8} - \frac{5}{8} =$ ☐

7 Start at 100, divide by 10, add 6, multiply by 2 and subtract 20. What is the answer?

8 What time is 20 minutes after 7:50 am?

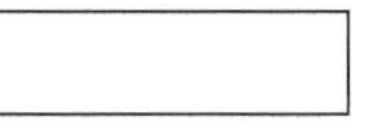

9 Draw **6 : 50** on the clock.

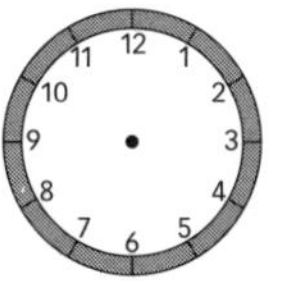

10 Circle the longest.

150 m 1 km 200 cm

11 Find the area.

A = ☐ m^2

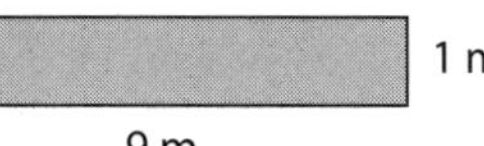

12 Write 476 g to the nearest 100 g.

13 How many axes of symmetry does the shape have?

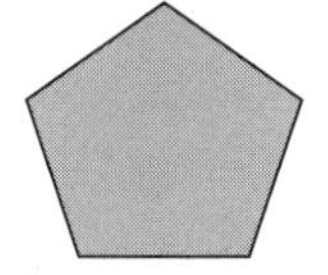

14 Name the shape.

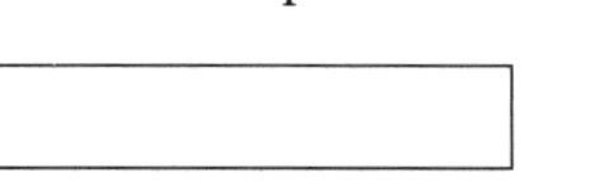

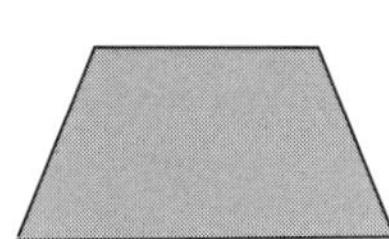

15 Draw a whale to the east of the island.

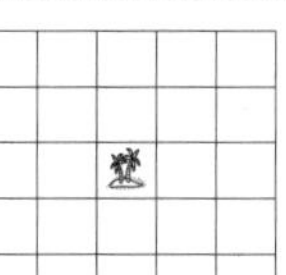

16 What is the distance, in units, between D4 and D5?

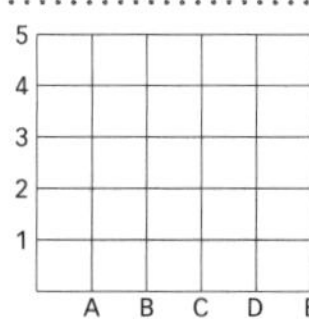

17 Record the size of the shaded angle.

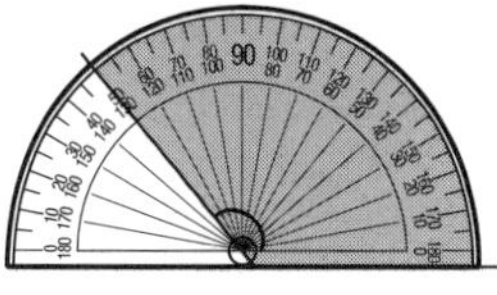

18 Give an example of an event that has a 0.1 chance of occurring.

19 What is the chance of selecting a spotty fish?

20 Complete the table of data using the graph.

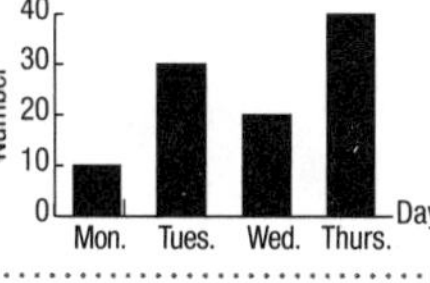

Day	Number
Monday	
Tuesday	
Wednesday	
Thursday	

21 What was the temperature at noon?

22 Draw a horizontal column graph for the data.

Colour	Number
red	50
green	10
white	65
grey	53
black	40

1

+	22	46	68	79	50	35
18						

2

–	93	102	41	65	80	119
40						

3

×	4	10	9	11	3	7
100						

4

×	1	6	12	5	8	2
9						

5

÷	700	600	900	500	100	300
100						

6

÷	15	30	45	20	50	25
5						

7 Complete the missing numbers.

$$\begin{array}{r} 4\ \square\ 6\ 3 \\ +\ 3\ 8\ \square\ 5 \\ \hline \square\ 9\ 0\ \square \\ \hline \end{array}$$

8 What is the total of 7296, 4381 and 4403?

☐

9

$$\begin{array}{r} 73\,000 \\ -\ 12\,468 \\ \hline \end{array}$$

10 What is 8000 – 1964 – 644?

☐

11 What are the first 5 multiples of 7?

☐

12 What is 10^2?

☐

13 $329 \div 10 = \square$

14 What are the factors of 48?

☐

15 What is the number of thousands in 163 192?

☐

16 Insert <, > or =.

100 000 + 40 000 + 10 + 600 ☐ 700 + 90 + 4000 + 10 000

17 Complete and round the answer to the nearest 5c.

$3\overline{)\$4.83}$ ☐

18 Complete the table.

Starting amount	Spent	Change
$50.00	$7.60	

19 $21 - 3.08 = \square$

20 Round 3.05 to the nearest tenth.

☐

21 Round $17.96 to the nearest dollar.

☐

22 $321 - 140 = \square + 56$

1

$$\begin{array}{r} 724\,366 \\ +\ 107\,955 \\ \hline \end{array}$$

2 What is the difference between 117 000 and 90 000?

3 How many seats are in a hall which has 9 rows of 56 seats?

4

$$8\overline{)3864}$$

5 Circle the smallest fraction.

$\frac{1}{3}$ $\frac{1}{5}$ $\frac{1}{4}$

6

$\frac{3}{10} + \frac{6}{10} =$ ☐

7 Start at 20, add 16, multiply by 3, subtract 23 and add 10. What is the answer?

8 What time is 10 minutes after 8:55 am?

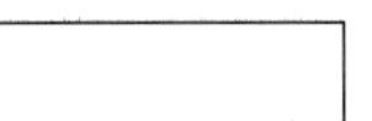

9 Draw **9 : 30** on the clock.

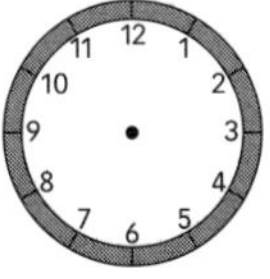

10 Circle the longest.

900 m 1 km 1000 cm

11 Find the area.

A = ☐ m^2

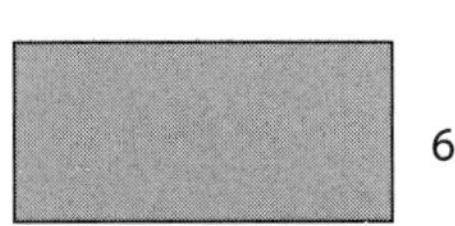

12 Write 1090 g to the nearest 100 g.

13 How many axes of symmetry does the shape have?

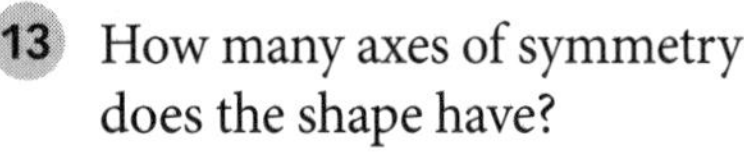

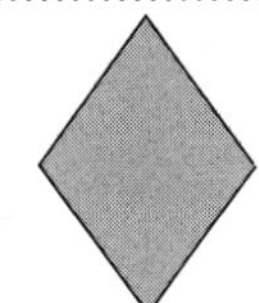

14 Name the shape.

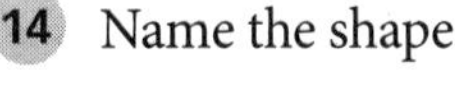

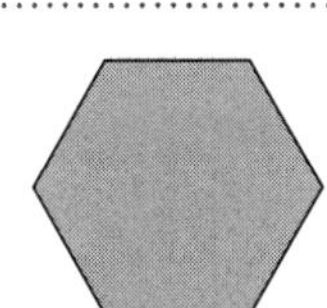

15 Draw a swimmer to the north-east of the island.

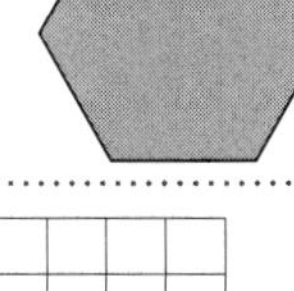

16 What is the distance, in units, between E1 and E5?

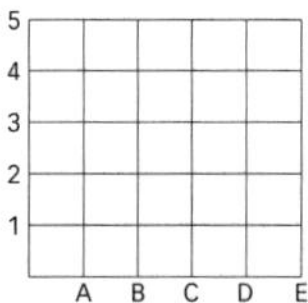

17 Record the size of the shaded angle.

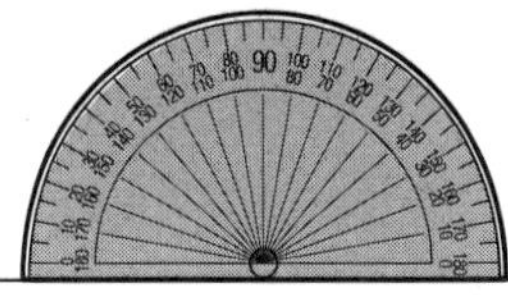

18 Give an example of an event that has a 0.4 chance of occurring.

19 What is the chance of selecting a plain fish?

20 Complete the table of data using the graph.

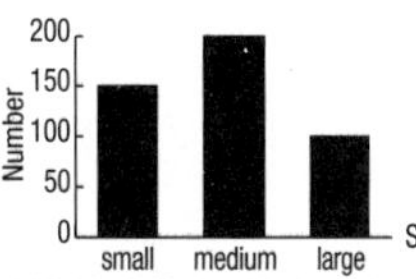

Size	Number
small	
medium	
large	

21 How many more students play basketball than football?

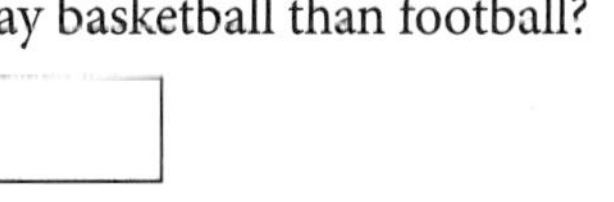

22 Draw a horizontal column graph for the data.

Book type	Number
fiction	20
non-fiction	45
poetry	12

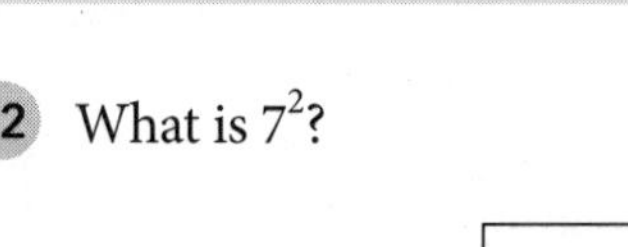

1

+	40	77	10	67	53	27
31						

2

–	50	75	49	84	58	36
25						

3

×	7	4	8	9	1	3
100						

4

×	5	10	11	0	6	2
3						

5

÷	1000	300	100	200	500	900
100						

6

÷	40	88	24	56	8	96
8						

7 Complete the missing numbers.

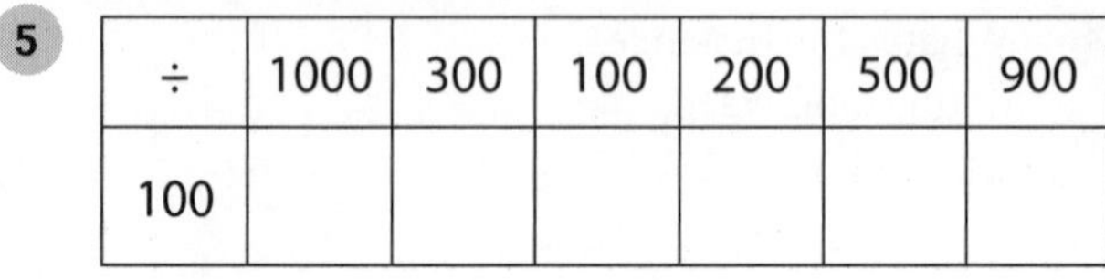

```
  □ 6 7 □
+ 4 □ □ 3
---------
  6 0 5 9
```

8 What is the total of 9846, 1124 and 3479?

9

```
  52 000
– 17 843
```

10 What is 7000 – 2946 – 107?

11 What are the first 5 multiples of 4?

12 What is 7^2?

13 $783 \div 10 = \square$

14 What are the factors of 45?

15 What is the number of thousands in 321 685?

16 Insert <, > or =.

3000 + 40 000 + 200 + 1 □ 900 + 5 + 50 000

17 Complete and round the answer to the nearest 5c.

```
  $22.50
– $19.99
```

18 Complete the table.

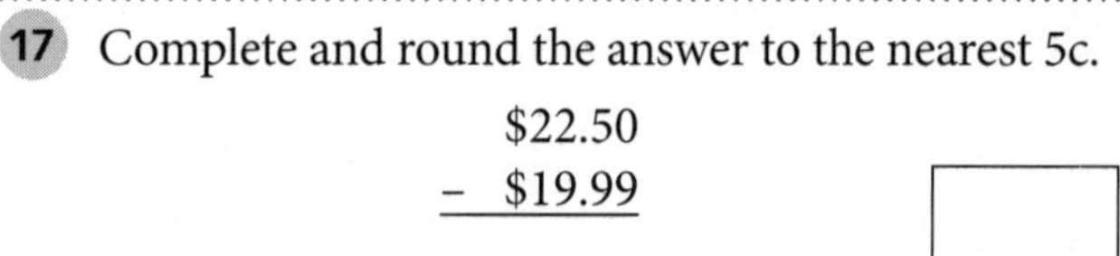

Starting amount	Spent	Change
$50.00	$14.05	

19 $25 - 17.63 = \square$

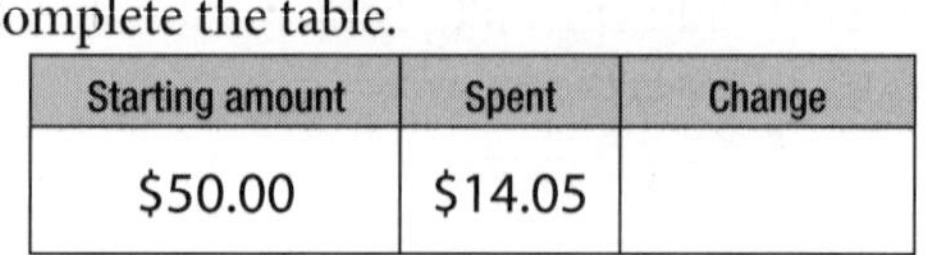

20 Round 7.99 to the nearest tenth.

21 Round $10.16 to the nearest dollar.

22 $326 + 95 = \square - 71$

1

$$\begin{array}{r} 699\,046 \\ +\ 324\,188 \\ \hline \end{array}$$

2 What is the difference between 111 000 and 66 000?

3 How many stickers are in 462 packets that have 9 stickers in each packet?

4

$$3\overline{)2001}$$

5 Circle the smallest fraction.

$\frac{1}{8}$ $\frac{1}{4}$ $\frac{1}{5}$

6

$\frac{7}{8} - \frac{1}{8} =$ ☐

7 Start at 200, divide by 2, subtract 35 and add 70. What is the answer?

8 What time is 16 minutes after 6:19 pm?

9 Draw  on the clock.

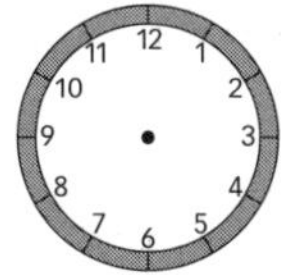

10 Circle the longest.

200 cm 10 mm 1 m

11 Find the area.

A = ☐ cm^2

12 Write 1360 g to the nearest 100 g.

13 How many axes of symmetry does the shape have?

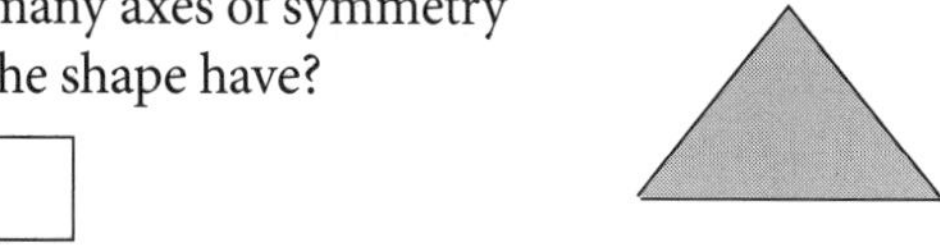

14 Name the shape.

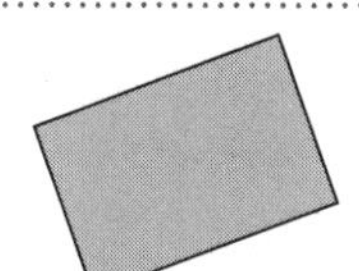

15 Draw a fish to the south-west of the island.

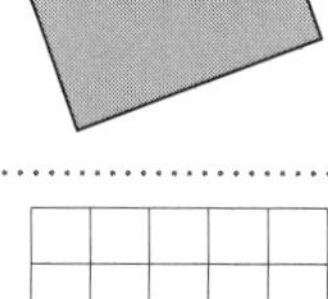

16 What is the distance, in units, between C0 and C3?

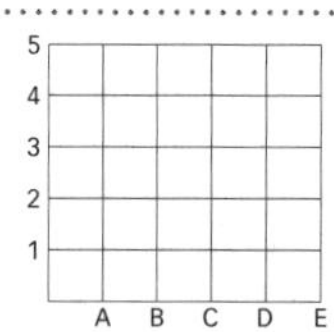

17 Record the size of the shaded angle.

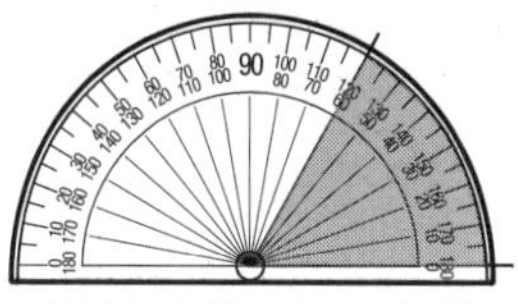

18 Give an example of an event that has a 0.9 chance of occurring.

19 Which fish has a fifty-fifty chance of being selected?

20 Complete the table of data using the graph.

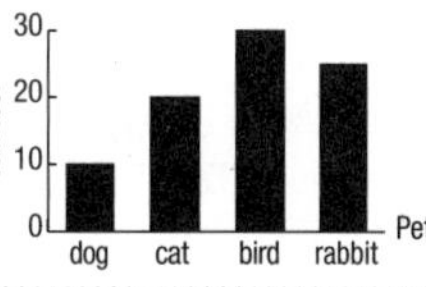

Pet	Number
dog	
cat	
bird	
rabbit	

21 What fraction of the garden is planted with roses?

22 Draw a horizontal column graph for the data.

Sea animal	Number
dolphins	20
sharks	5
whales	3
seals	10

1

+	29	18	53	67	7	46
12						

2

–	30	46	61	53	82	29
16						

3

×	7	5	9	4	11	3
100						

4

×	2	10	12	6	1	8
7						

5

÷	700	500	300	200	900	400
100						

6

÷	9	36	63	81	54	18
9						

7 Complete the missing numbers.

```
    5 □ 6 3
+   2 4 6 □
-----------
    □ 8 □ 1
```

8 What is the total of 3096, 1479 and 85?

☐

9

```
  96 000
– 72 463
```

10 What is 4000 – 1976 – 397?

☐

11 What are the first 5 multiples of 12?

☐

12 What is 5^2?

☐

13

$$912 \div 10 = \square$$

14 What are the factors of 70?

☐

15 What is the number of thousands in 1 216 439?

☐

16 Insert <, > or =.

20 000 + 200 + 80 + 1 ☐ 9 + 900 + 30 000 + 4000

17 Complete and round the answer to the nearest 5c.

```
  $2.53
×     4
```

☐

18 Complete the table.

Starting amount	Spent	Change
$50.00	$36.05	

19

$$13 - 6.03 = \square$$

20 Round 10.02 to the nearest tenth.

☐

21 Round $3.29 to the nearest dollar.

☐

22

$$421 + 76 = \square - 35$$

UNIT **30B**

1

$$\begin{array}{r} 486\,391 \\ +\ 216\,847 \\ \hline \end{array}$$

2 What is the difference between 86 000 and 111 000? [ ]

3 How many oranges are in 286 bags which have 8 oranges in each bag? [ ]

4

$$6\overline{)6192}$$

5 Circle the smallest fraction.

$\frac{1}{3}$ $\frac{1}{8}$ $\frac{1}{4}$

6 $\frac{5}{6} - \frac{2}{6} =$ [ ]

7 Start at 70, divide by 5, add 3, multiply by 2 and add 19. What is the answer? [ ]

8 What time is 25 minutes after 7:13 pm? [ ]

9 

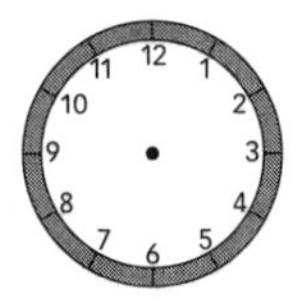

10 Circle the longest.

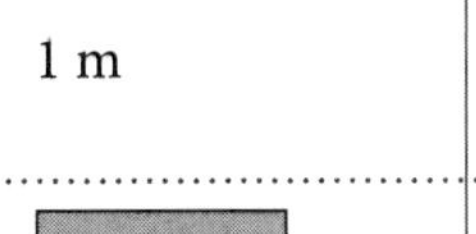

1 cm 200 mm 1 m

11 Find the area.

A = [ ] m^2

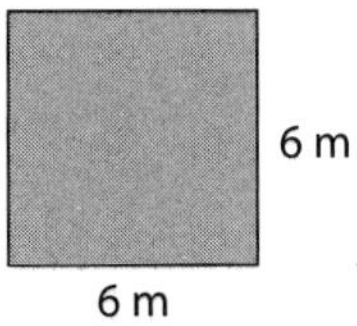

6 m
6 m

12 Write 990 g to the nearest 100 g. [ ]

13 How many axes of symmetry does the shape have? [ ]

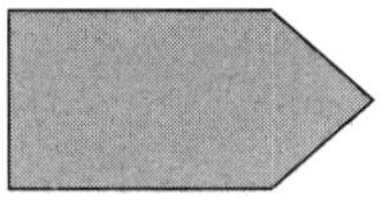

14 Name the shape. [ ]

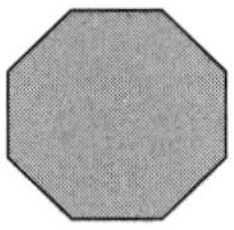

15 Draw a submarine to the south-east of the island.

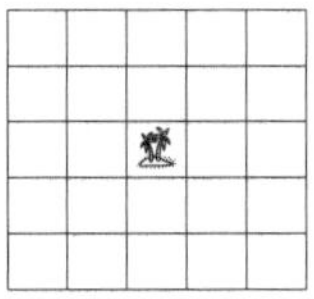

16 What is the distance, in units, between E3 and E5? [ ]

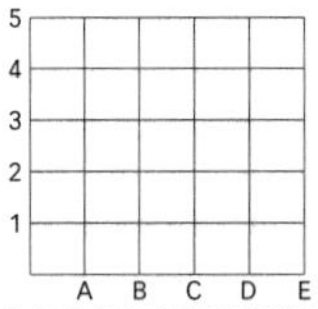

5
4
3
2
1
A B C D E

17 Record the size of the shaded angle. [ ]

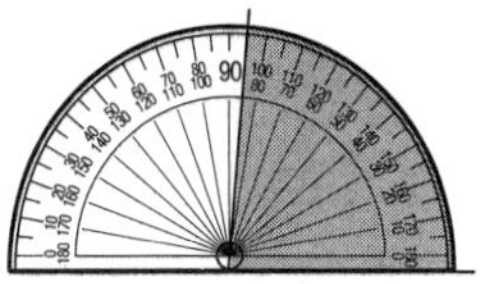

18 Give an example of an event that has a 100% chance of occurring. [ ]

19 Which fish has the greatest chance of being selected? [ ]

20 Complete the table of data using the graph.

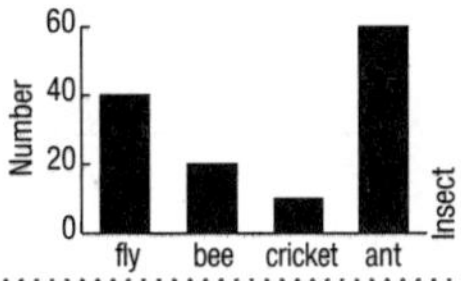

Number
60
40
20
0
fly bee cricket ant
Insect

Insect	Number
fly	
bee	
cricket	
ant	

21 What was the number of eggs laid at the farm on Wednesday? [ ]

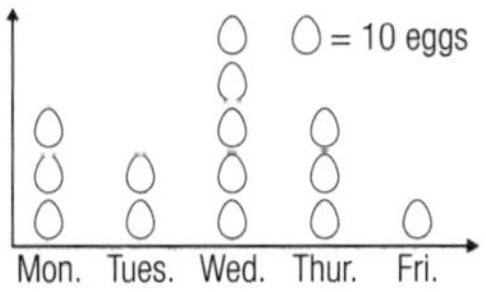

⬯ = 10 eggs

Mon. Tues. Wed. Thur. Fri.

22 Draw a horizontal column graph for the data.

Lunch	Number
sandwich	26
wrap	29
bread roll	13
sushi	17

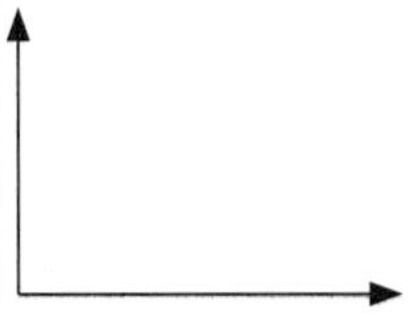

1

+	15	23	36	19	42	56
27						

2

−	29	67	48	50	71	89
18						

3

×	6	7	4	11	1	3
10						

4

×	2	8	5	10	12	9
6						

5

÷	400	900	1100	1000	300	700
100						

6

÷	84	70	35	42	14	7
7						

7 Find the missing numbers.

```
  4 7 □ 3
−   □ 7 9
---------
  □ 2 2 4
```

8 Add 6000, 90 000 and 2000.

9 Sarah has 1672. How much more does she need to make 9000?

10 1497 paint tins were sold from the 3500 at the warehouse. How many paint tins were left?

11 There are 6 sushi rolls in a packet. If there are 15 packets, how many individual sushi rolls are there?

12 What is 6^2?

A 6 **B** 12
C 36 **D** 66

13 What are the factors of 30?

14

327 ÷ 10 = □

327 ÷ 100 = □

15 What is the number of thousands in 613 207?

16 Insert <, > or =.

10 000 + 4000 + 90 + 3 □ 900 + 3000 + 60 + 1 + 10 000

17 Jack bought this top. What was Jack's change from $20?

$12.95

18 Each orange costs 60c. How many oranges can Aiden buy with $5.00?

A 1 **B** 5
C 8 **D** 9

19 6.27 + 3.09 =

A 9.36 **B** 9.39
C 10.17 **D** 12.07

20 Order the numbers from smallest to largest.

0.76, 0.67, 0.09, 0.60

21 Round $16.84 to the nearest 10c.

22 Complete:

+	20		27		15
19		65		51	

1 What is the sum of 593 117 and 247 854?

2 What is the difference between 50 000 and 700 000?

3 Find 7 groups of 356.

4 $4\overline{)6152}$

5 Which is the largest fraction?

A $\frac{1}{8}$ **B** $\frac{1}{4}$ **C** $\frac{1}{5}$ **D** $\frac{1}{3}$

6 Dad cooked a pizza. $\frac{6}{8}$ was eaten. How much of the pizza is left?

7 $50 + (9 \times 8) - 87 =$ ☐

8 What is the time 20 minutes after 7:45 pm?

A 7:55 pm **B** 8:00 pm
C 8:05 pm **D** 8:15 pm

9 Draw 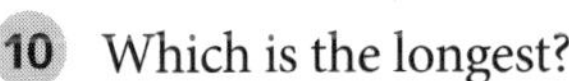 on the clock.

10 Which is the longest?

A 9000 cm **B** 96 m
C 9 km **D** 9 m

11 Find the area.

A = ________ m^2

4 m

8 m

12 Write 637 g to the nearest 10 g.

A 600 g **B** 630 g
C 640 g **D** 650 g

13 Draw the reflection of the shape.

line of reflection

14 Match the net to the object.

A **B** **C** **D**

15 Draw a bird to the north-west of the girl.

N

16 What is at (5, 3)?

17 Measure the size of the shaded angle.

18 List all of the outcomes for this spinner.

19 What is the chance of selecting a tennis ball from this bag of balls?

20 Complete the table of data for the graph.

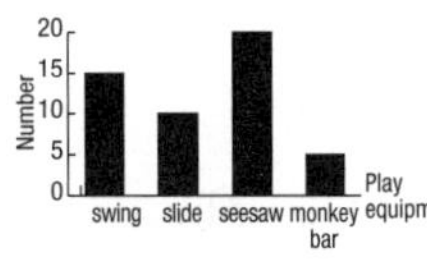

Equipment	Number
swing	
slide	
seesaw	
monkey bars	

21 What fraction of people went to the music concert?

22 How many more chicks than ducklings are there?

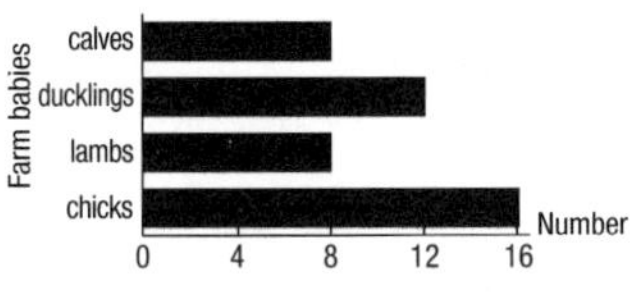

1 Alex has three cards.
He adds the cards together.
How much more does Alex need to make 100 000?

6000

400

70 000

A 34 600
B 24 600
C 33 600
D 23 600

2

$$90\,115 - 3179 = \square$$

3 Which number is closest to 3040?

A 4000
B 3000
C 3100
D 3020

4 Which has the largest value?

A 6^2
B $100 \div 10$
C $50 \div 2$
D 8×8

5 Four classes of 25 students travel to the pool by bus. Each bus holds 40 students.
What is the smallest number of buses required?

A 2
B 3
C 4
D 5

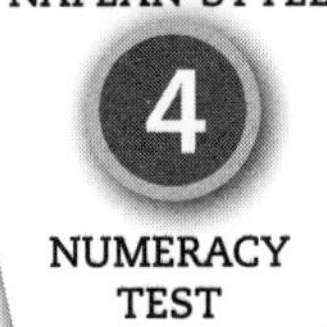

6

$$9640 \div \square = 964$$

A 0.1
B 10
C 100
D 1000

---

7 Jan has $28 and Lucas has $36.
How much does Lucas give Jan so they have the same amount of money?

A $2
B $4
C $6
D $8

---

8 7.64 is the same as:

A 7 + 0.6 + 0.4
B 7 + 6 + 4
C 7 + 0.6 + 0.04
D 7 + 0.4 + 6

---

9 A packet of three kiwifruit costs $2.
What is the largest number of kiwifruit that Kelly can buy for $87?

A 43
B 44
C 129
D 132

---

10 Here are 3 cakes.
They are shared equally between 15 children.
How much cake does each child receive?

A $\frac{1}{5}$ **B** $\frac{1}{3}$ **C** $\frac{1}{2}$ **D** $\frac{5}{1}$

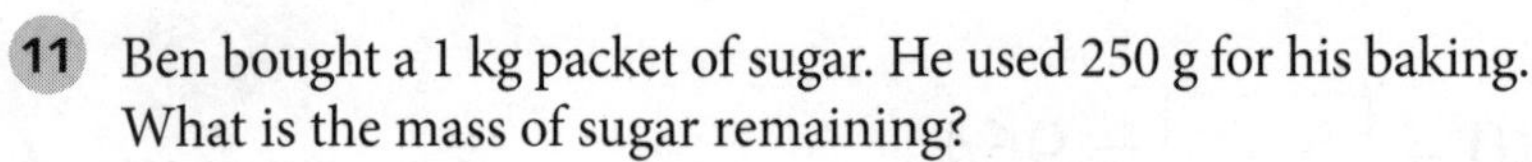

11 Ben bought a 1 kg packet of sugar. He used 250 g for his baking.
What is the mass of sugar remaining?

A 250 g
B 550 g
C 750 g
D 850 g

12 The school day starts at 8:50 am. Fruit break is at 10:00 am.
How long is the first session of the day?

A 50 min
B 60 min
C 70 min
D 80 min

13 The playground is a rectangle.
It is edged with logs.
Jill walks around the playground on the logs.
How far does Jill walk?

A 17 m
B 27 m
C 34 m
D 70 m

14 Which diagram shows a reflection?

A 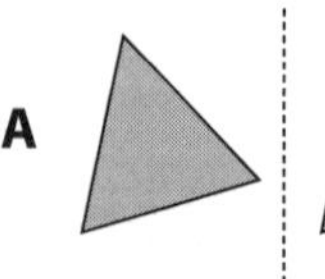

B 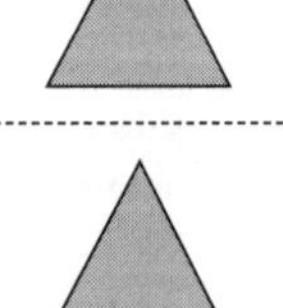

C 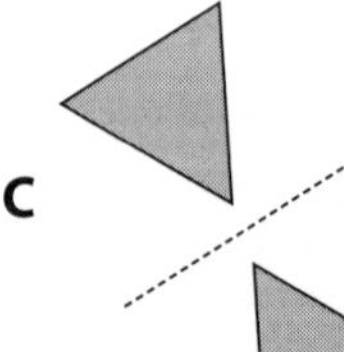

D 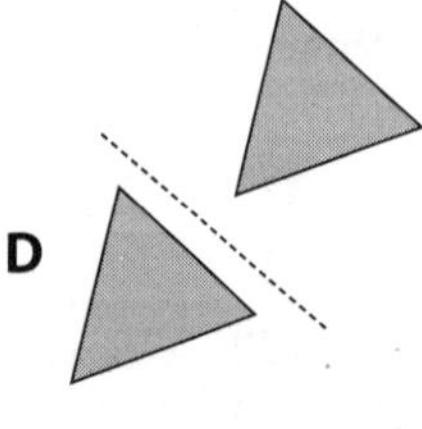

15 Here is a net.

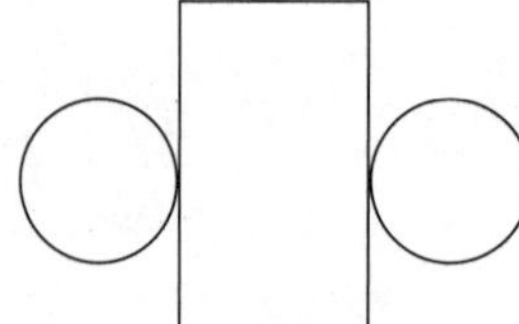

Which object is the net for?

A 

B 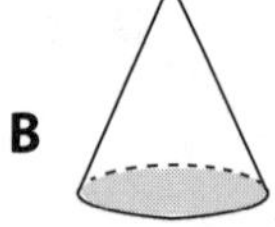

C 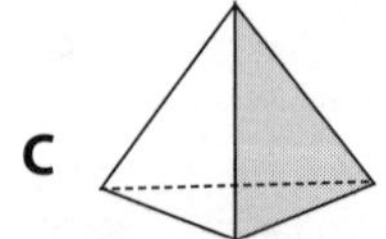

D 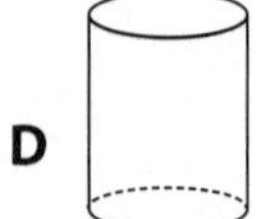

16 Jodie throws a six-sided dice to start a game.
What is the chance Jodie throws an odd number?

A $\frac{1}{2}$ **B** $\frac{1}{3}$ **C** $\frac{1}{5}$ **D** $\frac{1}{6}$

17 Sam measures an angle and finds it is a reflex angle.
Which angle could it be?

A 70°
B 110°
C 170°
D 300°

18 A survey was held to find the number of students who had a drink of water during the day at school.
How many students in Year 5 had drunk water at school?

Year	Boys	Girls
Year 3	15	12
Year 4	16	19
Year 5	17	15
Year 6	18	20

19 Here is the data for a car park over three days.
Which graph shows the results?

Day	Truck	Car	Motorbike
Friday	6	29	7
Saturday	7	20	3
Sunday	4	14	8

A 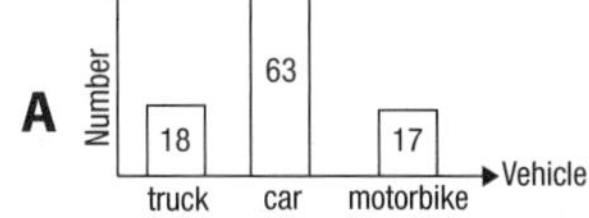

B 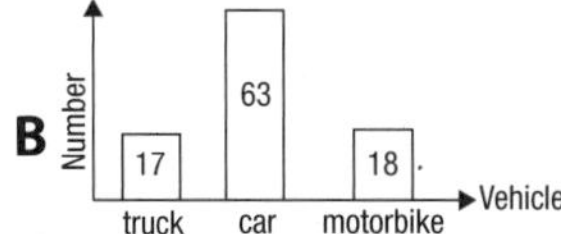

C 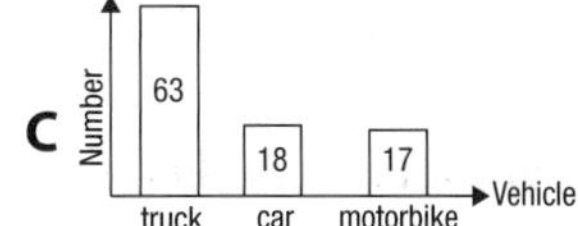

D 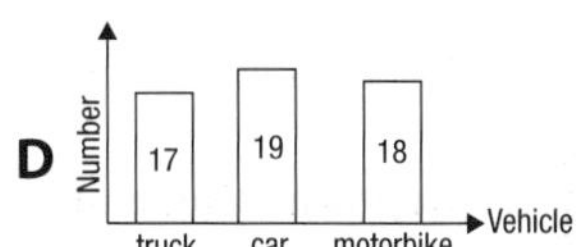

20 These models are made from identical cubes. Which model is made from 16 cubes?

A 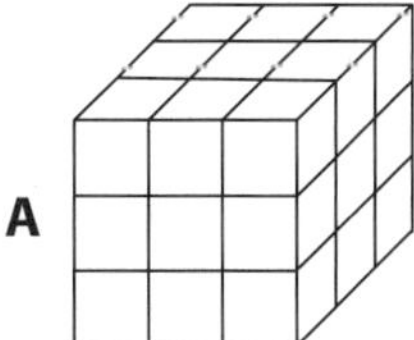

B 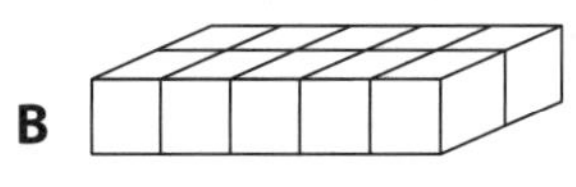

C 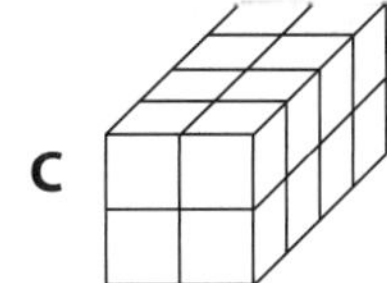

D 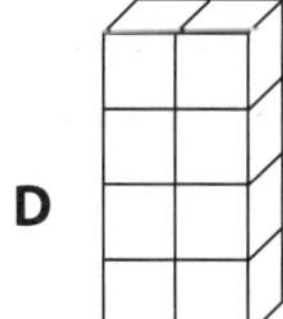

© 2018 Pascal Press
Reprinted 2020, 2021, 2022

Updated in 2023 for the NSW Curriculum and Australian Curriculum Version 9.0 changes

Reprinted 2024

ISBN 978 1 74125 620 8

Pascal Press
PO Box 250
Glebe NSW 2037
(02) 9198 1748
www.pascalpress.com.au

Publisher: Vivienne Joannou
Project editor: Rosemary Peers
Edited by Rosemary Peers
Answers checked by Peter Little
Page design by Kim Webber
Typeset by Grizzly Graphics (Leanne Richters)
Printed by Vivar Printing/Green Giant Press

Reproduction and communication for educational purposes
The Australian *Copyright Act 1968* (the Act) allows a maximum of one chapter or 10% of the pages of this work, whichever is the greater, to be reproduced and/or communicated by any educational institution for its educational purposes provided that the educational institution (or the body that administers it) has given a remuneration notice to Copyright Agency Limited (CAL) under the Act.

For details of the CAL licence for educational institutions contact:

Copyright Agency Limited
Level 12, 66 Goulburn Street
Sydney NSW 2000
Telephone: (02) 9394 7600
Facsimile: (02) 9394 7601
Email: memberservices@copyright.com.au

Reproduction and communication for other purposes
Except as permitted under the Act (for example, a fair dealing for the purposes of study, research, criticism or review) no part of this book may be reproduced, stored in a retrieval system, communicated or transmitted in any form or by any means without prior written permission. All inquiries should be made to the publisher at the address above.

The publisher thanks the Royal Australian Mint for granting permission to use Australian currency coin designs in this book.